MW01625819

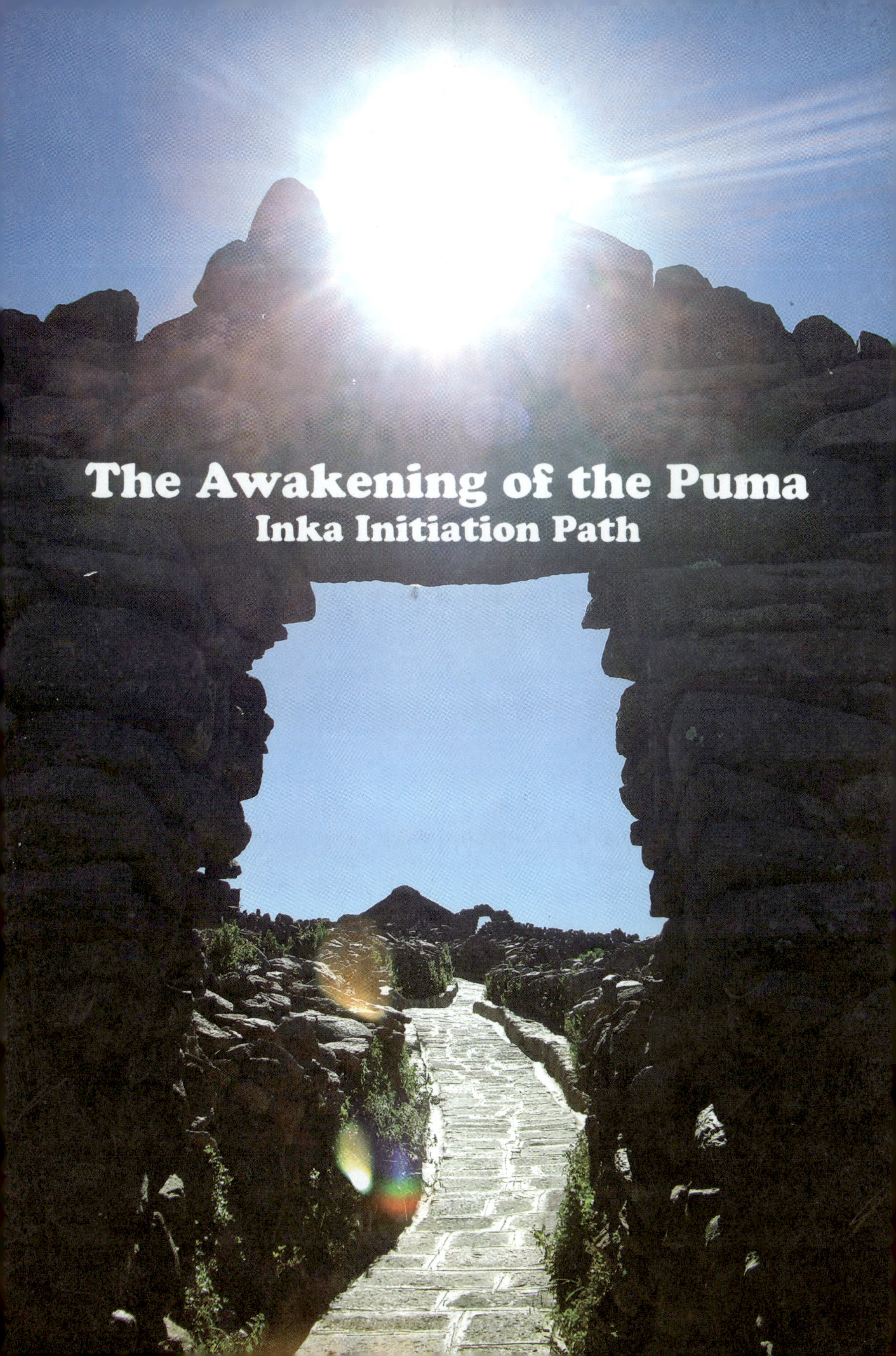
The Awakening of the Puma
Inka Initiation Path

The Awakening of the Puma
Inka Initiation Path
Evidences of Archaeo-astronomy in the Andes
Mallku Aribalo
Pachakamaq Shamanic Group S.A.C.
Group
Shamanic
Pachakamaq

Original title in Spanish:
"El Despertar del Puma, Camino Iniciático Inka"
(Evidencias arqueo-astronómicas en los Andes), 1997
By: James Arévalo Merejildo – Mallku
English translation by Beverly Nelson Elder
English correction by Helenita Ziegler and
Francisco Medina

The Awakening of the Puma
Inka Initiation Path
Evidences of Archaeo-astronomy in the Andes

Santa Catalina Ancha 366-B
Cusco - Peru
Phones: 0051/84/984761007 - 984760187 - 261419
E-mail: kondorpuma@gmail.com
Site: www.andes007.com

Tiraje: 2000 books

1st edition: Cusco - Peru 1997
2nd edition: Cusco - Peru 2010

Composition, production, photographic art, diagrams, drawings, maps, digital art and digital paintings: Mallku

Hecho el Depósito Legal en la Biblioteca Nacional del Perú N°: 2010-03124
ISBN: 978-612-45728-0-7

To my parents,
Gladys Sabina and Lucio with affection.

To my daughters
Jam-Aina,
Dariana-Sai and Dara-Taí.

To Pachamama, my mother and teacher,
and Wiraqocha, my guide.

ACKNOWLEDGMENTS

My special appreciation to those who helped with the English translation of "El Despertar del Puma."

To my spiritual brother Alain Goormaghtigh, for his suggestions.

To Elisabet Sahtouris, for the translation of the message from the wise man Asunción Acctu from Pachakamaq.

To professor *yachachiq* Teófila Vargas Salcedo for her direction in the *quechua* terms.

Everything in the Universe has its time and form; only the Heart of the World is forever. That is the Cosmic Law. Every epoch, like the great serpent Amaru, twists through eternity. In each of its turns, great power is hidden and the Unnamed One returns. In oblivion all thoughts are consumed, but the wise one learns the lesson: the Unknowable is not a single deity, but all of them are contained within its mystery. It always returns to speak in the places that give it voice, when someone is there to hear. In these places those who understand the celestial light... or those who listen to their own inner silence… will know. They are in the void that is the axis around which the unruly wheel of ignorance turns. At your father's side, when you saw the temple assaulted, a celestial Amaru bit its own tail. And the priests knew that in the ancient valley, where the voice of Irma lives, worship is called forth by the roar of the Earth uplifting the name of Pachakamaq*. Without knowing it, the sons of the Wiraqochas will name the god they thought they had destroyed. And the name is, "I AM." The time of your fulfilment is near. We have lived only for its arrival, to announce it to you.

(... teaching of Asunción Acctu, wise elder of Pachakamaq). "El Oro de Pachacamac" by Luis Enrique Tord.

The spelling of "Pachakamaq" is from us.

Nanasca, on June 22nd, about 20 to 30 centuries ago.

It was on the same day, June 22nd long ago, Sulka Nanasca, the old astronomer of the magnificent sacred city of Kawachi, slowly climbed to the top of an imposing pyramid in the middle of the desert. Soon, in the distance, the Sun would board the lighted raft of the night, surrounded by the mythical fish that would lead it to the quiet waves of Ni, the great sea to the South.

Sulka Nanasca, dressed in his long cotton tunic hemmed with hummingbird feathers, is intently observing the dusk. The high priest is a descendant of the noble lineage of the Long Heads who have been living in the plains between the Andean mountains and the Pacific Ocean for no less than five thousand years.

From the top of the pyramid, Nanasca is watching the nearly vertical fall of the solar disc that bathes the entire horizon in fire. If his calculations are correct, this day will be the shortest of the year and the Sun would disappear precisely at the end of a long, wide line of oxidized stones, which the priests who specialize in geometry have aligned in the direction of Ni, following his directions, as if an invisible giant had been leashed to the strap he holds in his hands. The Sun falls at once, precisely at the end of the line! Happiness floods the wise man's face. It almost seems as if the Sun followed the path traced with stones, chaining it to its nighttime resting place. At the foreseen hour, the Moon emerges precisely at the apex of the great triangle of Si, which Nanasca has drawn and starts to float above the spectacular desert valley.

The sky's animated spectacle fascinates the old man who presides over the destiny of the proud Nascas. With ecstatic voice, he speaks with each and every star, greets the planets and converses with the shining Southern Cross. He was right. The secrets of the cosmos belong to him! Tomorrow and forever, while the people of the sands go on living, other lines and other signs will record the perpetual movements of the Sun, the immense cosmos, and the unchanging path of the astral bodies which direct daily life and the well-being of his people.

(Selection from Simone Waisbard's book Les pistes de Nazca).

NOTICE

This work does not presume to detail the evolutionary process of the ancient Andean cultures. There are numerous books on this subject.

Instead, I'll be exploring little known aspects of the Andean archaeo-astronomy and magical practices of some ancient and Inkan cultures.

For a better comprehension of the native-Quechua expressions, see the glossary at the end of the book. For an easy reading of the Quechua words, the plural expressed with the suffix *"kuna"* is replaced by the "s" suffix as many authors are doing.

PROLOGUE

Mallku is a sensitive human being, a gifted Visionary, an Spiritual Leader and a warrior of light in this generation, with passion and devotion in promoting awareness, accuracy and clarity in all areas of research of Andean Culture. He is an internationally respected teacher, lecturer, prolific writer, photographer and humanitarian, revealing immense pride, dedication and expertise in presenting his amazing teachings and sharing of the Ancient Solar Initiations and the important secrets of Initiation and Shamanic experiences revealed by the Ancient Sages in their Sacred Power Places in the Andes. Mallku is an intuitive expert in the understanding and translation of the esoteric knowledge within Andean Archaeo-astronomy.

Human being now is turning his attention to the ancient lore and wisdom of nature discovering the meaning for his spiritual path. Some are driven by a desperate feeling that time is running out. Life is affected and too many species are disappearing, and too many people are losing their ability to live together in harmony and slowly realising that "Man did not weave the web of life; he is but a thread within it."

Mallku in "The Awakening of the Puma" brings us convincing messages of a time in Andean history when the masters in the Andes lived and celebrated with an intense conscious connection and interdependency with Pachamama - Mother Earth. With their foresight and respect to nature, these sages have left a tangible path clearly traceable for future generations.

Mallku guides our soul on an enlightening pilgrimage from power places and dimensional doorways, to healing springs, magic nature and celestial mountains. The physical journey through these highly power sites is a fascinating second step in this exploration to show us beauty and wisdom.

The land and people of this magical mountains of the Andes are indeed tremendously awe-inspiring, and combined with the addition of Mallku´s detailed research and clarity of explanation, we can witness the beauty of the past and present coming together, alive with the exciting promise of future continuity.

For the ones acquainted with shamanic experiences and practices of master cultures, some parts of these messages will motivate the recognition of their cosmic path and the need of uniting their voices for a better world. The initiate assimilates the secrets of nature, giving him intuition, dexterity, vision, and the ability to create and transform energy for the purpose of peace in human kind. One of the most interesting aspects of Mallku's teachings is that the doors to these new dimensions of awareness have recently been opened for us as a result of the beginning of a new time named "Pachakuti," or Cosmic Cycle. Andean history is divided into segments of one thousand years, with a transformational era after every five hundred years. In June solstice of 1992 we recently passed one of these historical transitions.

The visionary master Inka Pachakuti was responsible for designing the Andean city of Cusco in the physical shape of a Puma as bringing this celestial animal to Earth to extent illumination and strength to all inhabitants and pilgrims of all times. This Puma, center of the Inkan society, is progressively illuminated by each sunrise, symbolizing the gradual awakening of the transcendent development of awareness in human beings. Andean history was written by the invaders of a land which, after 500 years of domination, is now returning to a new epoch of balance and independence. Spiritual and psychic doors closed for centuries are now been opened, inviting and propelling us to new heights of responsibility and re-connection with our planet.

There is much more for us to rediscover about the foresight and vision of our ancestors. We are now in a new time, and the mission of this new epoch is related to, but different from that of the last cycle. Tolerance and respect between cultures, combined with the capacity to agree and disagree, is a major evolutionary step destined for the new Pachakuti. The ongoing and constant destructive fight for domination occurring between people of limited vision, who cling to and worship one manifestation of Divinity at the expense of another, will be replaced by recognition of the many aspects of the divine.

Mallku announces that other secrets of the Andean wisdom will be explored in subsequent books. Subjects on the healing practices and magic connections of the coastal areas and jungles. The ceremonial use of power plants, and the interaction with sacred sites in cosmic time and the guidance to the amazing world of the Andean masters with the intention to awake our own Solar Puma and make peace come true on Earth.

Keep in mind that the mere fact of just being in these sacred places will restart the natural balance with the other realms of nature. Your different bodies will experience and reconnect themselves with the inner forces of our Mother Earth/Pachamama, who is giving us one more chance to bring our consciousness into alert for the different events about to come.

Catherine Harvey,
Reiki Teacher, Healer and Writer.

Honoring Pachamama.

PRESENTATION

Time exists, where does time go?

In our lives we ask ourselves a lot of questions and some of them often remain unanswered. What are the dimensions of the Universe? Where does space begin and where does it end? I won't try to give answers to these questions because as man has walked this world, many of his concepts have undergone changes and the definitions given about various aspects of life and the Universe have continually been modified. When Europeans and other non-native peoples arrived in America, the erroneous concept generally accepted in their Christian world that "the Earth was flat and full of monsters and deep chasms," collapsed. Columbus was not the first to get to America, but he gave proof in his time that the Earth was round and inhabited by creatures who were not descendants of Adam. We see in our history how, on a daily basis, deeds and new evidence change the concepts and definitions of our surroundings and ourselves. There is no culture or group of people who owns the absolute truth about the many stages of human beings and their natural surroundings.

Educators, teachers and masters have come, always repeating the saying "Know thyself and thou shalt know the Universe." To know the Universe means "to know only one truth and its multiplicity."

During these past two thousand years, we have seen the development of dominant cultures which in turn imposed their influence on others, often to the point of extinction, thus giving way to other peoples' development in an unending evolution. It's only in this last century, at the end of the Christian era, that man, believing he holds the key that opens all doors to the understanding of the enigmas of life, has limited his field of investigation.

Modern man has decided that something must be tested in a laboratory in order for it to be understood. This uses only part of our integral being, such as our senses of sight, touch, hearing, smell and taste, forgetting that man is a spiritual being. That is to say, he is linked to an unknown Universe where a human being has other undefined aspects, possessing many perceptions, which are not recognizable and therefore rejected by the greater majority.

Thousands of years ago, it was known that the Earth was round; that a year contained approximately 365 days; and that the Earth was not the center of the Universe, but only a point within it. The same was true for the eternal principle that says "nothing is created and nothing is destroyed, it only eternally transforms itself." Something that has a beginning cannot be eternal and how can something begin if it has no end? This is why, in the ascended initiatory centers, the masters conceived the integrity of the being, defining it as "as above so below" and if what is above is infinite and eternal, so, also, that which is below. Therefore human being is also infinite and eternal.

Our rational mind must lead us to our goal and our reasoning tells us that there isn't just one path, but many; and that beyond the point, there is the circle. Absolute white or black does not exist, they are the product of all tonalities in between; man and woman are not different from one another, but share many traits. A house has more than just one function; it is not only to live in, but also to protect one from the surroundings of cold, heat, wind, rain, animals and other men, etc.

We ask ourselves how pre-Columbian and other great cultures that have walked the earth, sometimes had precise knowledge of astronomical manifestations, of planetary and stellar dimensions, of humankind and the Universe, without also possessing all the technology that we have today? We know that they had vast knowledge that allowed them to build cities with a solar orientation, to know about the geographic and magnetic North, to verify the precession of the equinoxes, the arrival of the solstices, the passage of the Sun at its zenith, the inclination of the axis of the Earth, etc. How could we believe that these people only used their five senses?

The most recent scientific reports confirm that the Earth of long ago did not have the same rotational movement that we experience today and that the angle of inclination of its axis was also different. This shows us that thousands and millions of years ago, the length of a day and a night as well as the differences between the seasons was not the same as today. We also had a very different geographical landscape that included many animals and plants now extinct.

I would like to present this investigation as one more page in the great book of life. The sole purpose is to awaken our Inner Sun and make us more conscious of the fact that we possess an age-old culture and mystery which, due to its complexity and depth, can never be completely resolved or understood.

The Author.

PART I

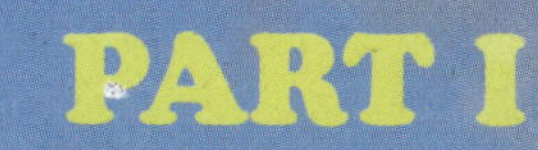

PACHA, THE UNIVERSE, THE EARTH AND HUMAN BEING

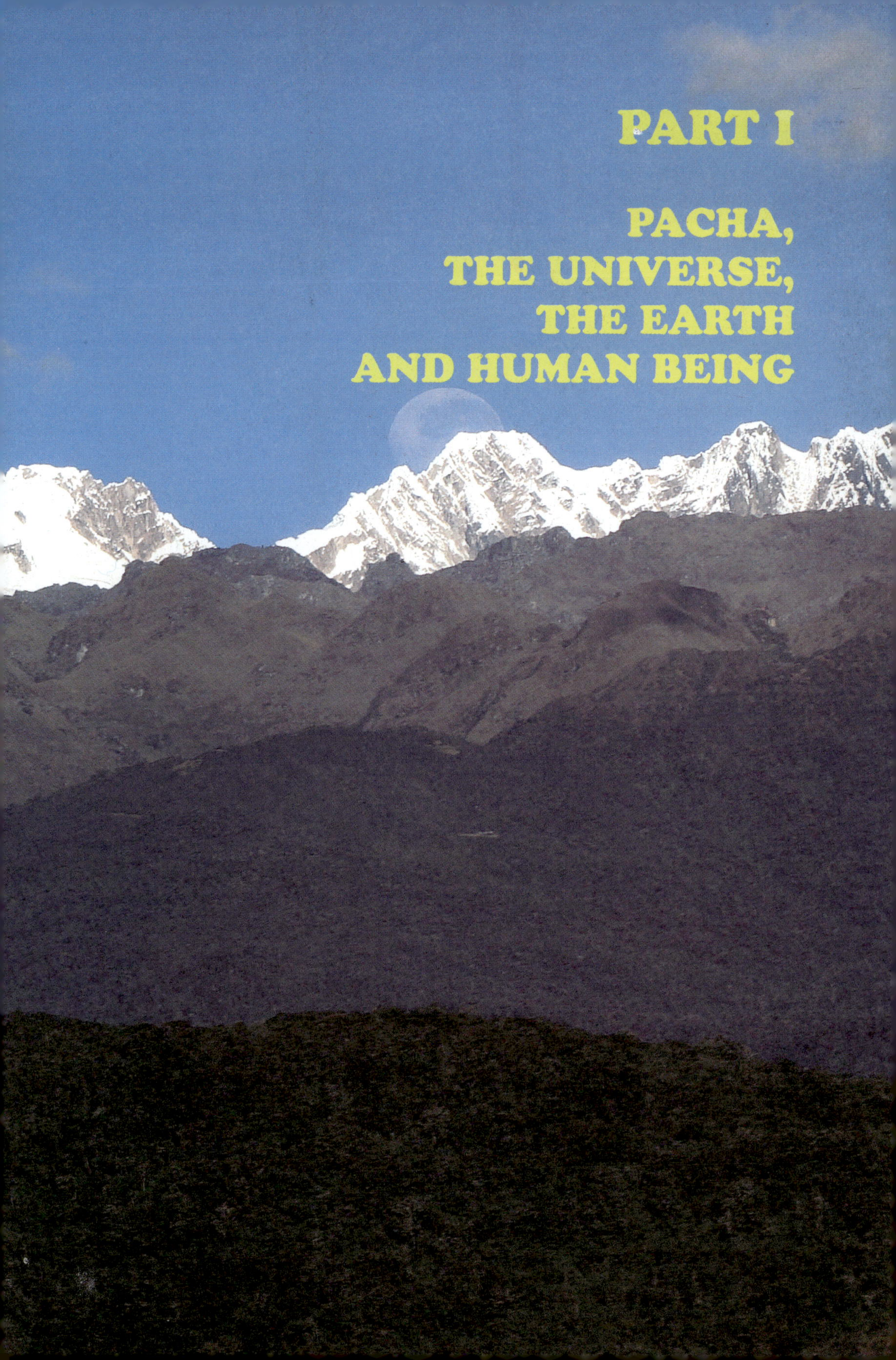

1. ABOUT HUMAN BEING, THE EARTH AND THE SUN ELEMENTS OF ASTRONOMY

Man* is defined as Homo Sapiens, in other words, the intelligent and thinking human being who attempts to define himself and the Universe. It's very hard to try to put the phenomenon called life under a microscope to define it, especially when life presents us with such an infinity of interdependencies where everything is linked together, with each man's life but a thread in a great universal web.

Actions and reactions are taking place in every moment of our lives. From dawn till when daylight disappears, we undergo multiple changes and are subject to incomprehensible influences. Nature directs our movements, we are dependent on her and our apparent decisions are subject to thousands of planetary and cosmic factors. For example, we are in error if we say, "It's freezing cold!," because a short distance from us, we'll see someone else dressed in only a bathing suit or short sleeve shirt. This shows us how geographical latitudes and differences in altitude affect our attitudes. It's the climate and temperature changes that make us create habits and customs. Little do we care if these are contradictory to the habits and customs of other human groups.

Observing life in a superficial way, we might think that we are acting independently from everything and that we are making our own decisions, asserting that one does one thing or another because he wishes it so. The illusion of free choice has made us create laws and precepts, accepted by our societies. It seems that we'll never be short of words when we talk about these philosophical things. Since the dawn of humanity in every nation, some of our world leaders and great thinkers have asked themselves the same questions. Who are we? Where do we come from? Where are we going?

There is a Taoist saying that goes: "Human being follows the world's law, the world follows the heavens' law, heaven follows the cosmic law, the cosmos follows the Tao's law and the Tao follows its own law." Perhaps this might be the answer!

* When I use the term man, I refer to humankind in general.

From the moment when man realizes that after day comes night, and once again daytime returns, the idea of time as a measure takes shape within his mind. He notices that sometimes he gets up very early in the summer and very late in the winter; that sometimes the Sun is visible, at other times the sky is cloudy, that some days are rainy and others snowy. He also finds that some places are humid and others dry, that he breathes freely at sea level and with difficulty when on a high mountain. But one thing that for sure, is that after each night comes a new day. So whether we like it or not, our lives are surrounded by permanent cosmic influences. These inescapable influences have created different responses from mankind through the eons, but always with the result of helping humans adapt and survive in their environment.

In the morning, if we look to the East, we'll see the Sun rise and if in the evening we look to the West, we'll see the Sun go down, thus creating the phenomenon of day and night. But what causes it?, is the rotation of our planet. What is interesting is that we see the Sun rise and fall at different points of the horizon as the days go by. We might say that the Sun is moving within certain limits depending upon the observer's geographical position.

Each day and night are the result of the continuous rotation of the Earth, and it is calculated that our planet needs approximately 24 hours to complete one rotation around its axis, a cycle called day. The different seasons that we experience, are the result of the inclination of the Earth's axis with respect to the elliptical orbit of the earth taken in its annual trip around the Sun.

Dear friends, I'd like to mention that the work I am presenting does not pretend to be of extreme scientific rigor. I'll only demonstrate fundamental elements for a simple and reasonable understanding of some of the astronomical phenomena that take place in our world.

MOVEMENTS OF THE EARTH

Everything is in motion, everything vibrates (Hermetic axiom).

The Earth needs approximately 365 days, 5 hours, 48 minutes and 46 seconds to go around the Sun, to complete a cycle called a year. In our calendar, the year is measured as 365 days; this is why an adjustment must be made every four years. Because the leap years are not sufficient to complete the Solar Year, some adaptations have to be made at the end of each century. The year 2000 A.D. will be a leap year during which the annual time of the Twentieth century will be completed (This book was published in 1997).

Day and night are the result of the rotation of the Earth around its axis in a time period of 23 hours, 56 minutes and 4 seconds, which we call sidereal day. The Earth's path around the Sun isn't exactly circular, but has an elliptical shape, where we are at times closer or farther away from the Sun. The closest distance from the Sun, called perihelion, is of approximately 146.4 million km. This is taking place when it's summer in the Southern hemisphere. During the aphelion, which is the farthest distance between the Earth and the Sun, we'll be about 151.2 million km apart.

The result of the orbital movement of the Earth around the Sun is the regular repetition of the seasons; taking into consideration that the distance between the Earth and the Sun is not a determining factor in the differences that exist between the seasons. Let's remember that the seasons are a consequence of the inclination of the Earth's rotation axis with respect to its orbital plane, this inclination angle is of 23°27′. This inclination will allow the ecliptic, which is the plane on which the Earth's orbit traces its elliptical trajectory around the Sun, to cut the Equator's plane obliquely, also forming an angle of 23°27′. It is this oblique angle which makes it possible for four points to exist within the ecliptic: two solstices and two equinoxes.

INCLINATION OF THE EARTH'S AXIS

The Earth's axis is inclined according to the elliptical orbital plane of the Earth, at 23°27′.

SOLSTICES

When the North Pole is inclined towards the Sun, it's summer in the Northern Hemisphere and at the same time, it's winter in the Southern Hemisphere. This is the aphelion, which also is the maximum distance separating the Earth from the Sun. After six months, the South Pole will be inclined towards the Sun, thus corresponding to summer in the Southern Hemisphere and winter in the Northern. In this case, we are in perihelion, which is the shortest distance between the Earth and the Sun.

There are winter and summer solstices. In the Southern Hemisphere, the winter solstice is on June 21st and the summer solstice on December 22nd and in the Northern it will be the opposite. The 21st of June is the day when the Sun passes over the tropic of Cancer, which for the Northern Hemisphere, is the day of summer solstice. In the Southern Hemisphere, the day the Sun passes above the tropic of Capricorn corresponds to the 22nd of December, the day of the summer solstice. In accordance with the geographical latitude and the weather conditions, the royal Sun culminates in its highest point or midday when the solar heat reaches its maximum intensity.

For the observer, the Sun does not always rise in the same place.

IN THE SOUTHERN HEMISPHERE

1. The Sun rises in the East with a tendency to the South (December 22nd).

2. The Sun rises in the East and sets in the West (March 21st and September 23rd).

3. The Sun rises in the East with a tendency to the North (June 21st).

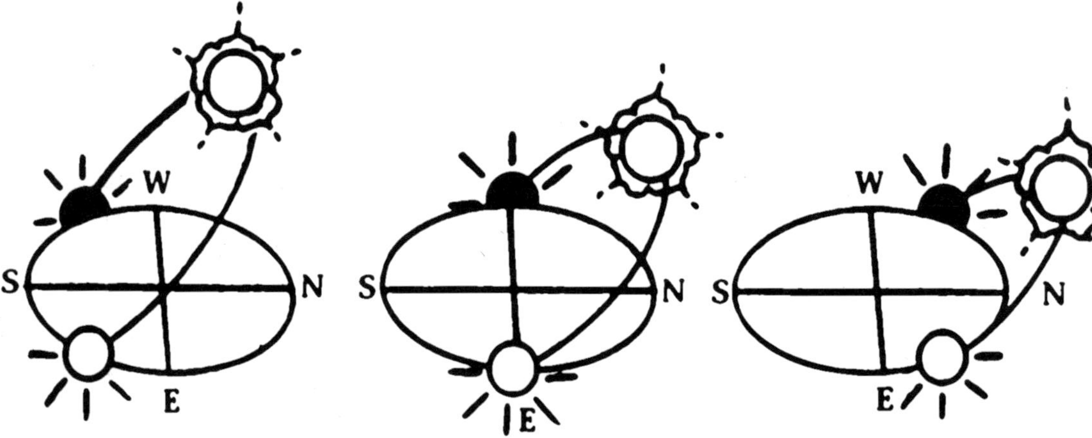

EQUINOXES

These are the days of the year when the Sun apparently crosses the celestial equator, intercepting the ecliptic. The result of this movement is that the days and nights will have equal duration, as long as this is in accordance with the observer's geographical surroundings.

In the Southern Hemisphere, the 21st of March is called the autumn equinox and the 23rd of September, the spring equinox. For the Northern Hemisphere, it will be the opposite. So let's say in a simple way that the Sun rises exactly in the East on those days (21st of March and 23rd of September) and will also set exactly in the West.

The Sun does not rise and set in the same place every day. There are variations in these movements, which we can see when the Sun is in the Southern Hemisphere. It rises in the East on the 21st of June, with a tendency toward the North and a tendency toward the South on the 22nd of December.

THE SEASONS FOR THE SOUTHERN HEMISPHERE

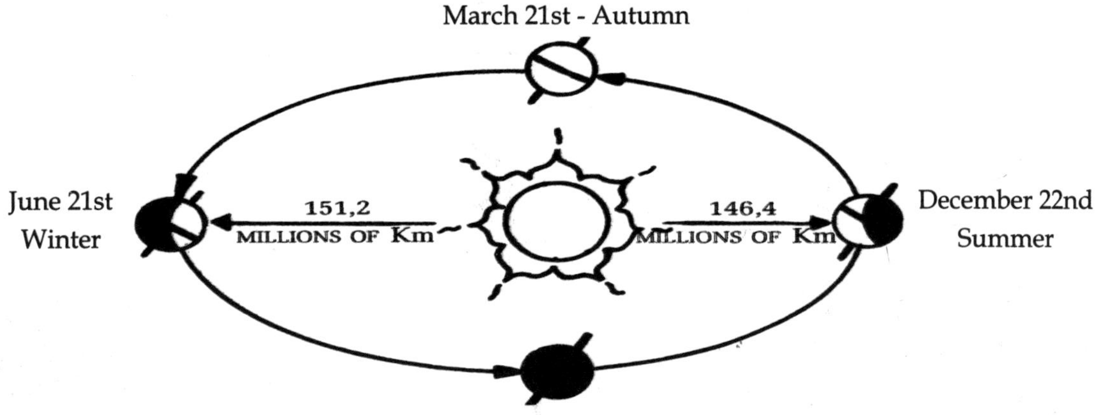

*The Earth follows an elliptical orbit around the Sun. Its closest point in relation to the Sun will be on December 22nd. That minimum distance is of 146.4 millions of km (perihelion). On June 21st, the maximum distance (aphelion) of about 151.2 millions of km is reached.

*The distance between the Earth and the Sun, does not determine the existence of the seasons, since these result from the inclination of the Earth's axis, in accordance with its movement around the Sun.

THE GREGORIAN CALENDAR

Today, we are using the Gregorian calendar that evolved from the Roman calendar, which is based on the Earth's annual movements around the Sun.

In order to allow the days to be counted in round numbers, Julius Caesar decreed that there were 365 days in a year, and to compensate for the annual loss of a quarter day, he created the leap years of 366 days every four years.

These calculations were not precise, because the time it takes the Earth to go around the Sun does not correspond to 365 days and a quarter, but to 365 days, 5 hours, 48 minutes and 45,98 seconds. This figure gives us a displacement of three days every four hundred years.

In 1582 Pope Gregory XIII decided to correct this displacement by eliminating one leap year every four centuries (those that end with two zeros and cannot be divided by four hundred). This readjustment determines that the years 1700, 1800 and 1900 are not leap years, but the years 1600 and 2000 are.

The Sun remains our principle indicator for measurements of time. We know that the duration of the days has not always been the same. There have been changes in the past and we can be sure that there will be more in the future.

REFERENCE OF THE RISING OF THE SUN

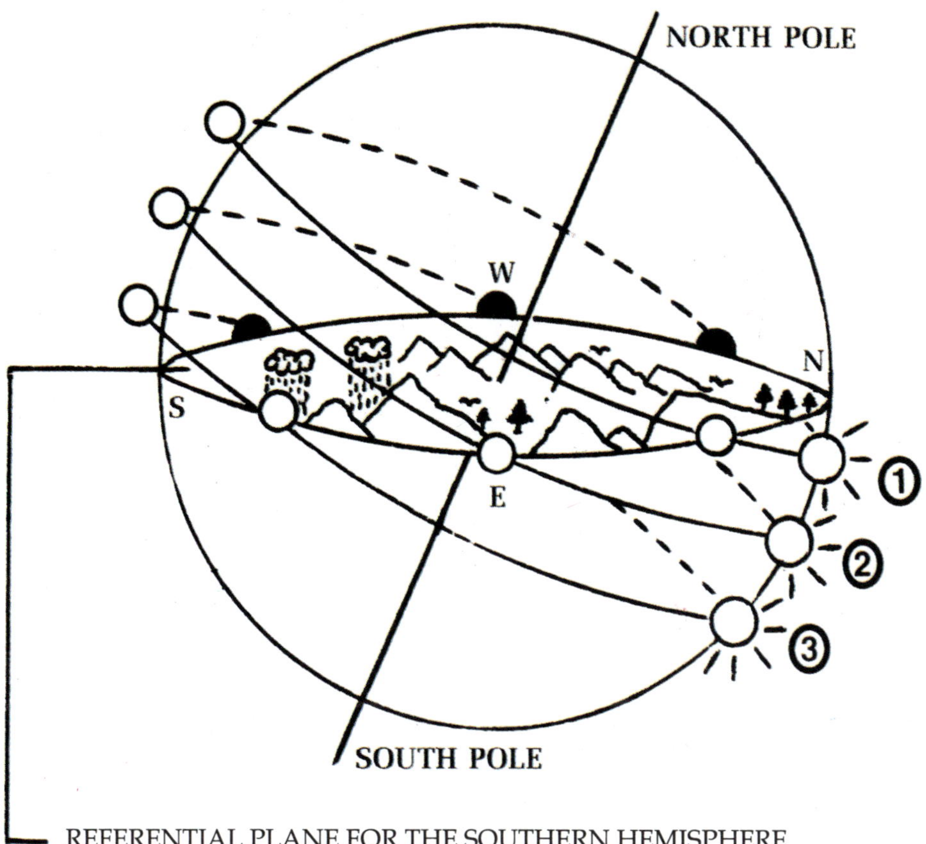

REFERENTIAL PLANE FOR THE SOUTHERN HEMISPHERE

1. Path taken by the Sun during June* solstice.
2. Path taken by the Sun during the equinoxes (September 23rd and March 21st).
3. Path taken by the Sun during December solstice.

*When observing the sunrise and sunset we'll see that the Sun rises exactly from the East on the days of the equinoxes and will set exactly to the West.

*Now, during the solstices, we'll see differences in the sunrise and sunset. The differences will take place to the North and to the South of the East or West points of the horizon.

*On December 22nd, we'll observe the greatest difference to the South and on June 21st, the greatest difference to the North.

*Note: For respect to both hemispheres, I will not refer any more to summer/ winter solstice or autumn/spring equinox. Instead I will use the terms: June solstice, December solstice and March equinox, September equinox.

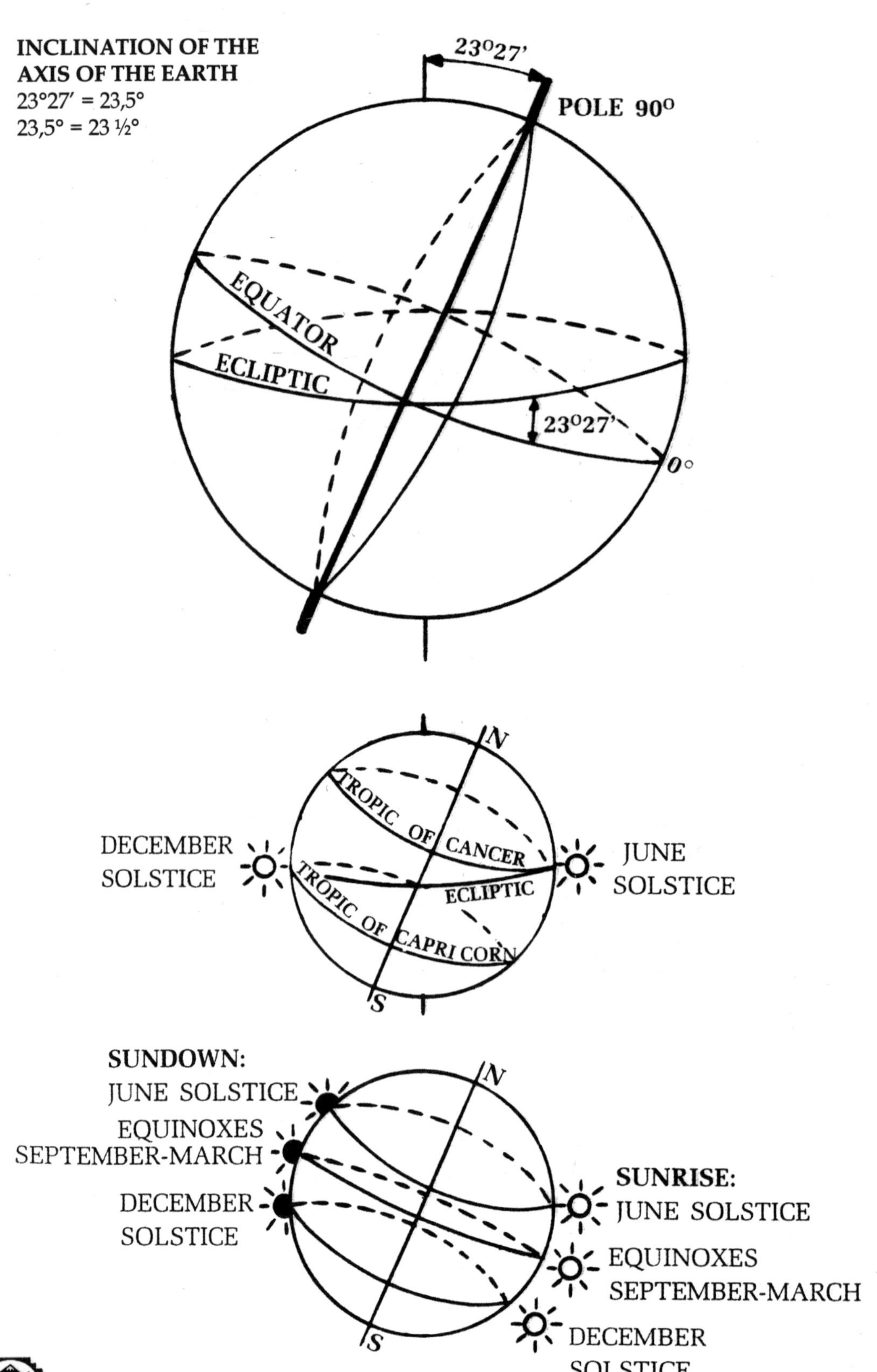
INCLINATION OF THE
AXIS OF THE EARTH
23°27′ = 23,5°
23,5° = 23 ½°
23°27'
POLE 90°
EQUATOR
ECLIPTIC
23°27'
0°
N
TROPIC OF CANCER
DECEMBER
SOLSTICE
JUNE
SOLSTICE
ECLIPTIC
TROPIC OF CAPRI CORN
S
SUNDOWN:
JUNE SOLSTICE
EQUINOXES
SEPTEMBER-MARCH
DECEMBER
SOLSTICE
N
SUNRISE:
JUNE SOLSTICE
EQUINOXES
SEPTEMBER-MARCH
DECEMBER
SOLSTICE
S

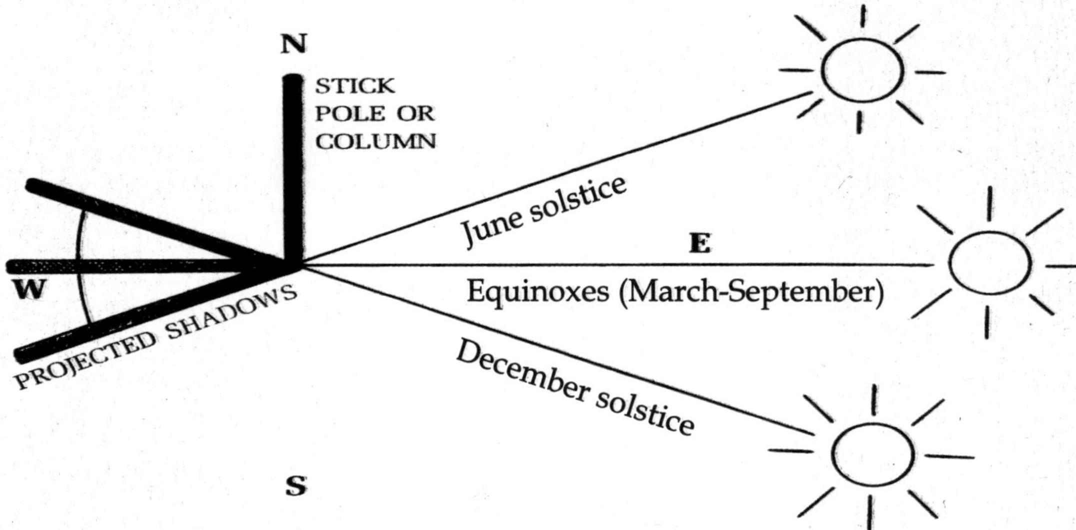

Elementary system for following the Sun and marking the seasons. Since long ago, the solstices and the equinoxes were well studied. (Model from the book "La Civilizacion Andina" by Hugo Boero Rojo).

W

WANKA

S

N

E

December solstice

June solstice

(September and March equinoxes).

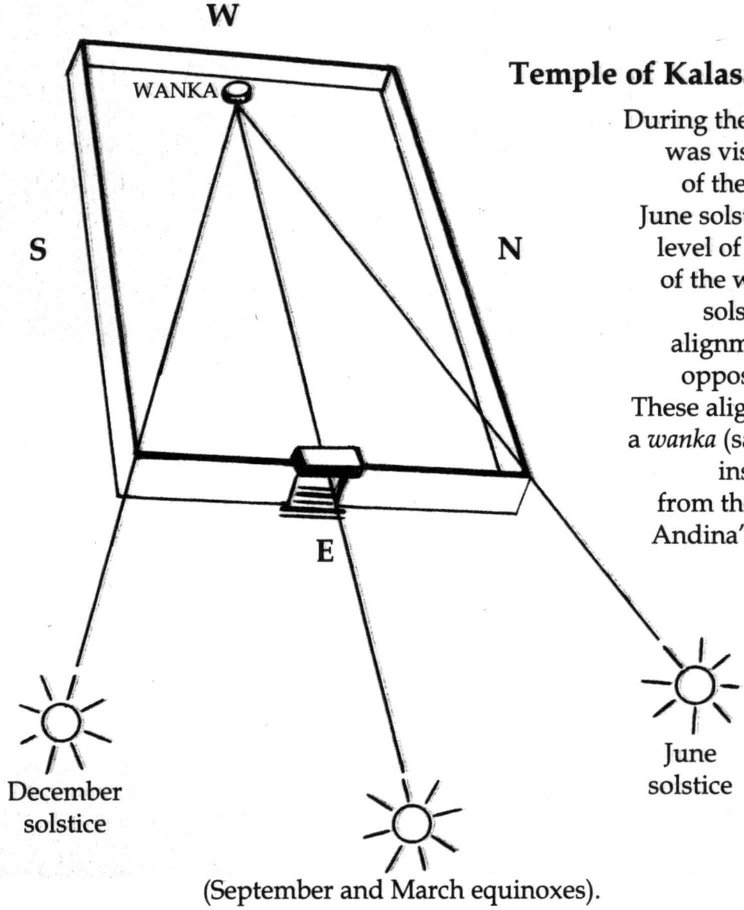

Temple of Kalasasaya in Tiwanaku

During the equinoxes, the sunrise was visible through the center of the main door. During the June solstice, the Sun rises at the level of the Northeastern angle of the wall. And the December solstice was marked by the alignment of the Sun with the opposite Southeastern angle. These alignments were related to a *wanka* (sacred rock of reference) inside the temple. (Model from the book "La Civilizacion Andina" by Hugo Boero Rojo).

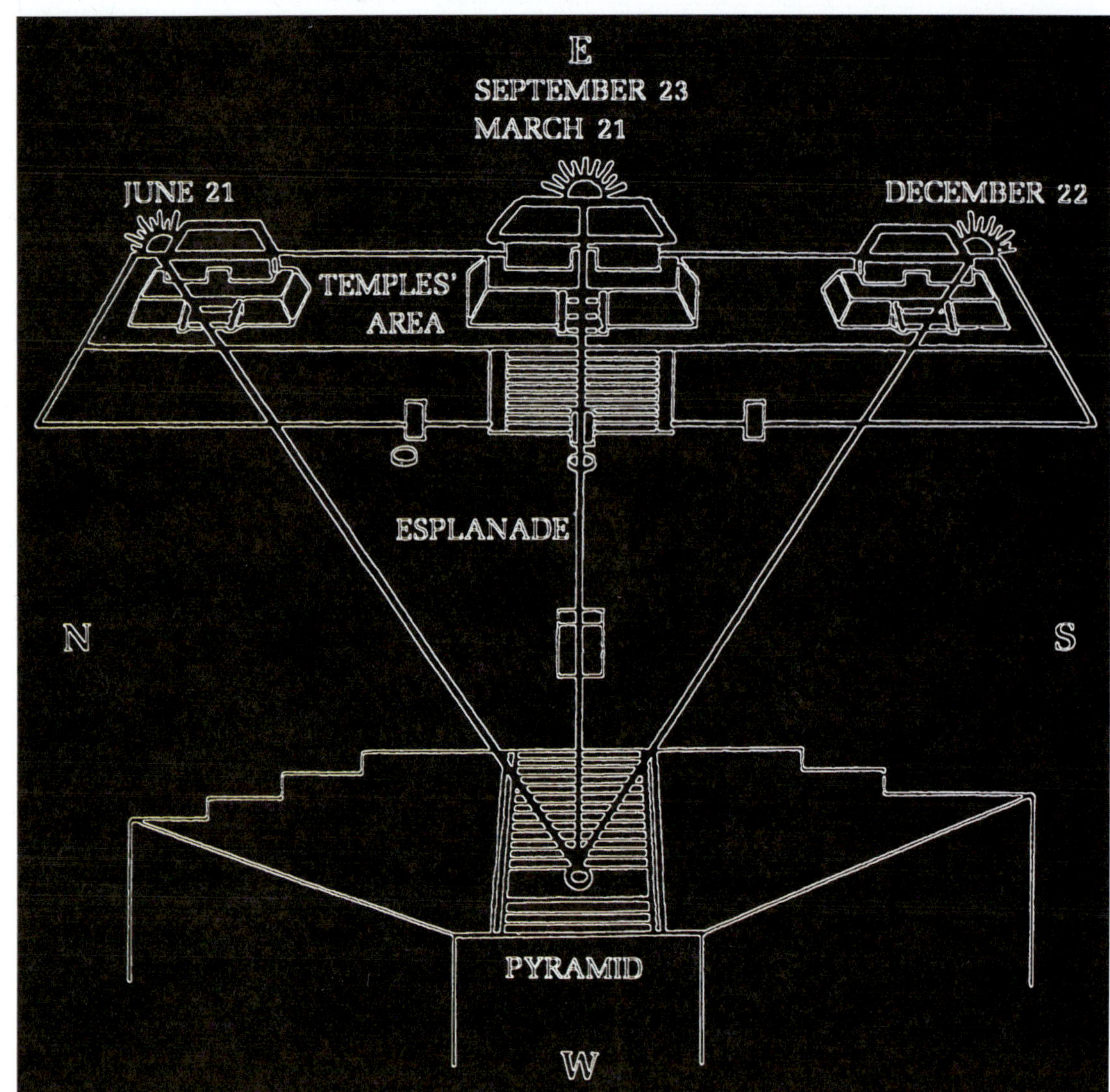

Mayan solar observatory of Uaxactun in Peten (Guatemala). In the old Mayan society, astronomical observations were also made to determine the solstices and equinoxes. Seen from the observation point on these dates, the Sun appears to rise from behind the temples (Model from the book "La Ciencia Antigua y los Zodiacos del Viejo Mundo" by Dick Edgar Ibarra Grasso).

2. EARTH: BLUE PLANET, OUR MOTHER

"Everything that happens to the Earth
will also happen to the children of the Earth"
(Indian Chief Seattle).

Our Earth, blue and unknown planet, is the Father and Mother of our first parents, sustainer of our races. The Earth is our Mother, our greatest entity and guide of our paths.

We can't look at life from the outside, because it presents itself before our very eyes, in the first breath we take into our bodies and the small breath already within us. Life is the most fascinating and yet most mysterious experience. It wasn't subjected to scrutiny until man observed and asked, as the wonder pulsed within him. These questions, sooner or later would become the same for every conscious entity. Who are we? Where do we come from? Where are we going?

The most ancient teachings of humanity have always guided us to the essence of Being, inviting us to revive what sleeps within us, the primordial movement that generates life. The wise always said that everything is in motion, that everything vibrates... everything is alive!

If everything is in motion and vibrating, then life is present everywhere in a wide variety of colours and manifestations. Life is not exclusively humankind's, but belongs to everything that surrounds us, everything we can touch, see, feel, as well as many things we do not see that are beyond our immediate reality. Life encompasses everything. Our planet is the main support system for everything that lives on it. The great Indian Chief Seattle gave his answer to the white man in this clear message.

...Teach your children what we have taught ours;
that the Earth is our mother.
Because everything that will happen to the Earth,
will also happen to the children of the Earth,
If men spit on the ground, it is on themselves that they are spitting.
Man did not weave the web of life; he is but a thread within it...

We are the sons and daughters of this Earth, she is our Mother, she is the one who keeps us alive, she is the closest living entity to us and she shelters us. Human being represents the microcosm of this great macrocosm which is our planet. Our blue world is similar to us; it also has organs, vital centers, polarities, positive and negative energy charges, meridians, energy centers, life cycles, and other manifestations, which for most of us, are beyond common perception.

Human being is an ontological being, and is linked to everything, just like our Earth. The brain and the heart are the centers from which our life expresses itself, thus representing an energetic unity of the positive and negative poles necessary for the unity's harmony, in this case, from human nature.

Our planet, which does not differ from us, also has its electromagnetic centers situated in precise regions of the globe. These centers are the Great Himalayan mountain range and the Andes, representing the positive and negative poles, which the masters and sages defined as the centers of masculine and feminine magnetic radiation. Certainly until this century, the Himalayan pilgrimage centers, temples and monasteries, sheltered almost a hundred percent of the masculine beings. This is not the result of mere chance, but simply because of the type of energy that will continue to emanate from the Himalaya, basically masculine. The complement of the Himalayan masculine energy is the Mother's energy, which emanates from the Andes and the center of this feminine pole is located at Sacred Lake Titikaka. This feminine energy moved the Andean people's consciousness so deeply that them identified this Earth as their Mother, calling her Pachamama.

Most of the world sees spirituality coming from the Orient. Today when we look for guidance such as meditation, health and natural medicine, we direct our attention to the Orient, because until recently, that culture has been made accessible to the Western world. For us, the children of Pachamama, this spiritual life develops in the Andes and their surroundings. But this ancient wisdom, still found in the mountains, valleys, plains and deserts, was existing for a whole cosmic cycle called Pachakuti. This cycle, which was completed just a few years ago, produced the beginning of a great revolution in the collective consciousness that will certainly modify the very life structures of our generation. This consciousness of the new humanity who is stepping into the third millennium, is the consciousness of the Andean chief and leader who said, "I'm going; but when I come back, I'll be millions."

The complement of the great Oriental culture whose guardians were the sacred mountains of the Himalaya (in Sanskrit, Home of the snow) is in the Americas, the Andes mountain range, which is the largest in the world.

The shape of the Himalayan mountain range looks like a slightly curved half-moon with a total length of approximately 2415 km and a width fluctuating between 160 and 240 km. The highest summit of the world is in that mountain range, at 8848 m above sea level, Mount Everest, name given in honor of Sir George Everest for his expeditions between 1830 and 1843. It was recognized as the highest peak in the world in 1852 and baptized Everest in 1865. The sherpas or snow tigers always called the peak Sagarmatha with great reverence, which means Goddess of the Universe, and the Chinese people saluted it with the name Qomolangma.

The Andes' mountain range is the largest in the world. This volcanic mountain range has an approximate length of 7500 km and its greatest width is 600 to 700 km from the southern part of Peru and the middle of Bolivia. The highest peak of this range is Aqonkawa, at 6959 m above sea level. The Quechua people call it Akqon-kawak which means Stone Sentinel and the Mapuche people knew it as Aqon-we indicating, that which comes from the other side.

The Andean people always revered the mountains, because they knew about the electromagnetic power that resided in them. They called them Apus. The Apus are the great entities which inhabit the summits of the mountains.

The Andes.

The Andes.

Our planet's electromagnetic centers are situated in two precise areas: the Himalayan and the Andean mountain ranges. Their respective highest summits are the Sagarmatha or Qomolangma known as Mount Everest at 8848 m above sea level, and the Aqonkawa at 6959 m above sea level.

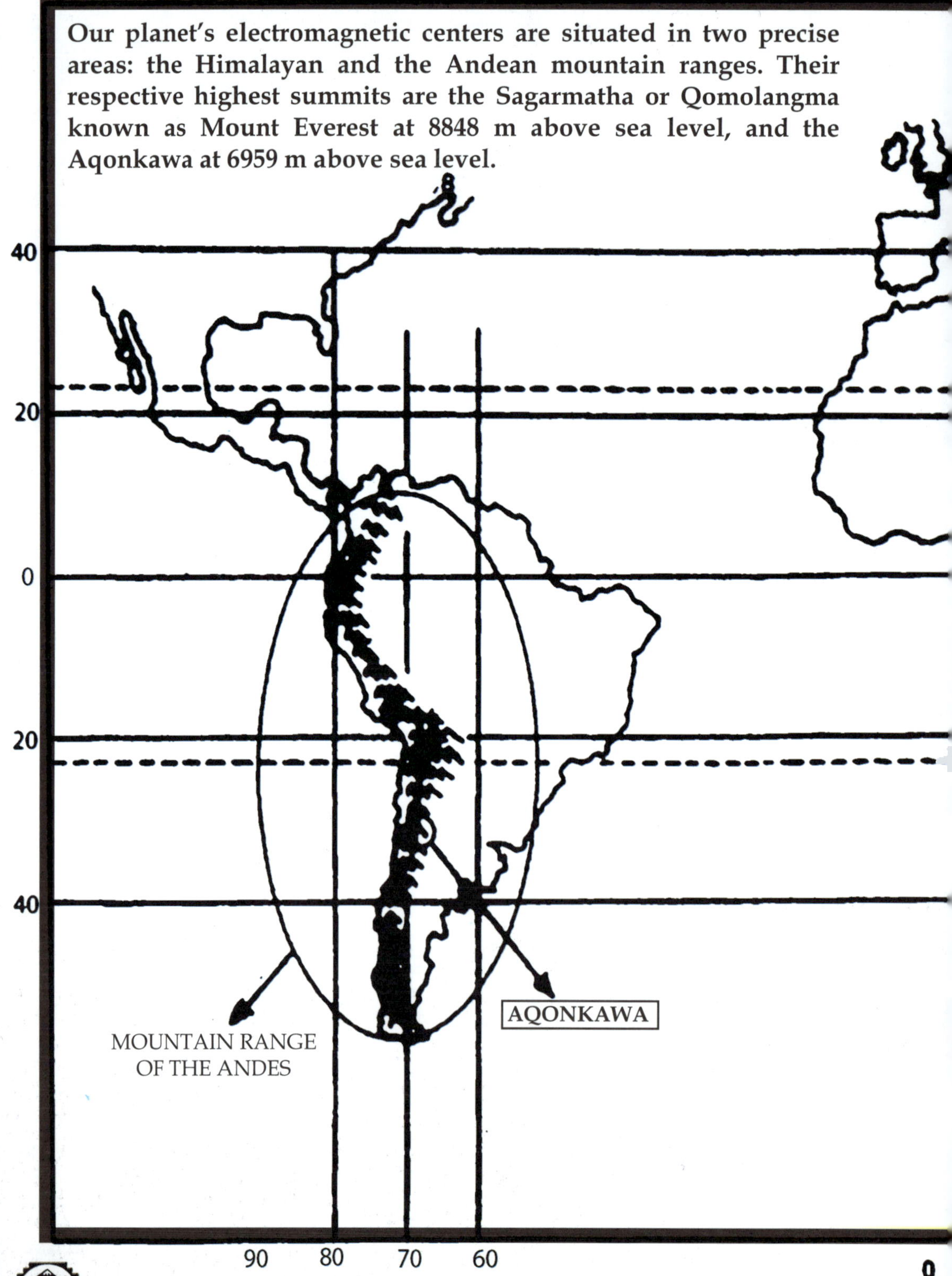

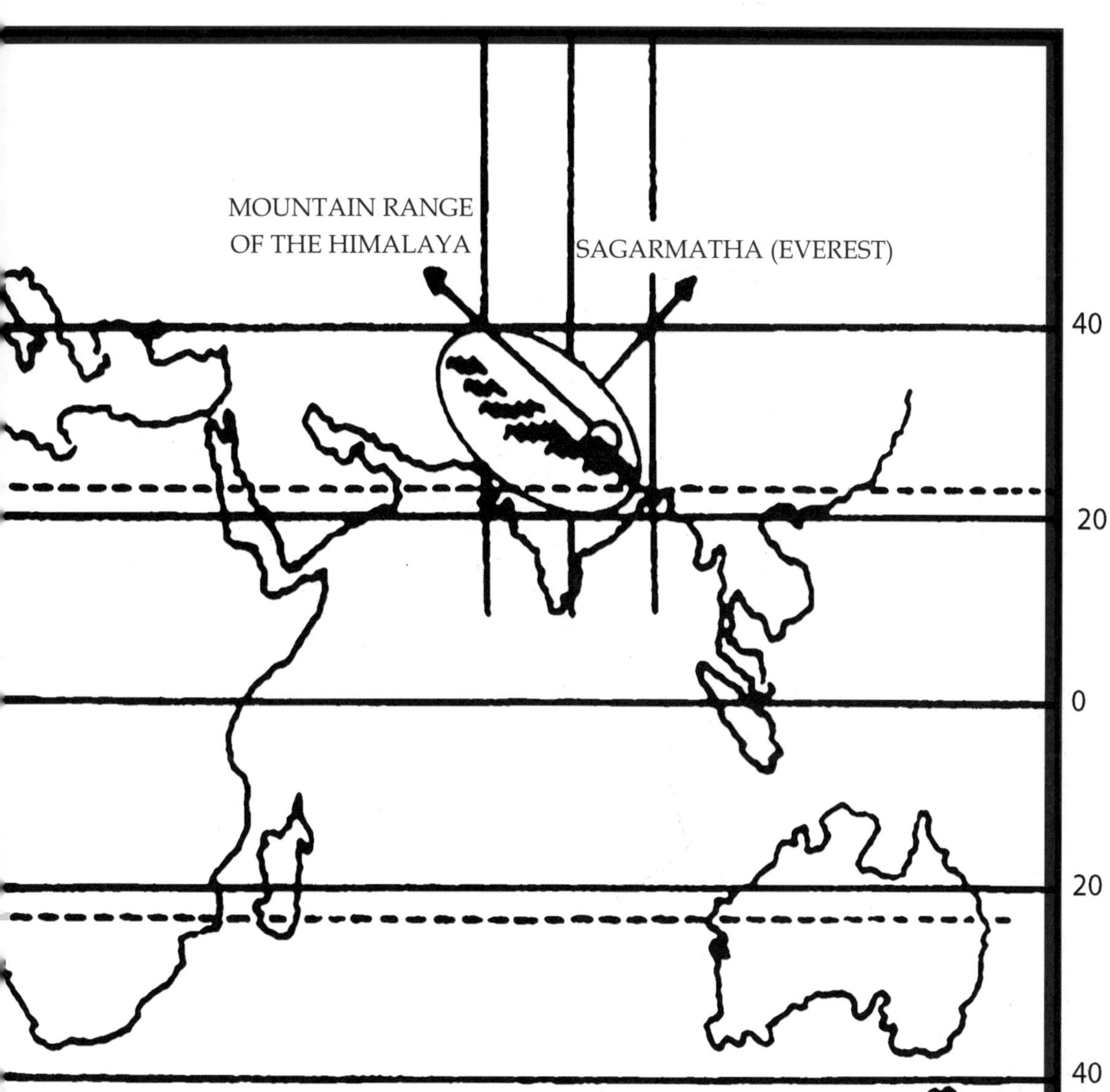

SAGARMATHA and AQONKAWA are the highest peaks in their respective hemispheres. One represents the eternal vigilance of the Orient, and the other, the guardian of the Americas.

The geographical position of both mountains is astounding. Both are situated between latitudes of 20° and 40° (North and South) above and below the Tropics. The Himalaya follows longitudes between 70° and 90° (North-East), and the Andes are situated in longitudes of 60° to 80° (South-West).

Meditation in the Andes.

3. THE ANDEAN PEOPLE, CHILDREN OF THE SUN

There was a time when everything was good.
A happy time when our mother took care of us.
(Sayings of the Nahuatl Sages).

According to modern knowledge, man appeared in America between 30 and 40,000 B.C., using different ways to migrate by land and sea. For the history of the Earth, a difference of 10,000 years in calculating the appearance of man in America really isn't much to speak about, but for man's earthly life, it is enough time for hundreds of generations. Many theories are being questioned and it's certain that someday we'll be confronted with astounding new discoveries, such as the realities that quantum physics show us year after year. But until otherwise demonstrated, we'll continue to use the small clues that we now hold in our hands.

To talk about the Andean people is also to speak of the ones who live in the geographical areas around the Andes, between the Pacific coast and the foothills of the mountains and in the tropical rain forest. Once more, I'll be referencing the experts who study these Andean cultures. They've studied diverse human groups with complex societies who continued to forge the higher cultures finally producing the Children of the Sun.

The wonderful part about these Andean cultures is that in their birth, growth and dying processes they never suffered a radical rupture, or completely disappeared. On the contrary, the essence of their knowledge was preserved through the masters and initiates and transmitted to the ones ready to receive it. Parts of this knowledge and practice are still alive today, though they are now being adapted to this century's rhythms and types of expression. I assume here that the inner wisdom was not completely destroyed or repressed.

The best known cultures of the last millennia, existing in different times and geographical contexts, have been baptized with new names: Sechin, Moxeque, Chavin, Kupisniqe, Kaluyo, Parakas, Tiwanaku, Naska, Mochika, Pachakamaq, Wari, Chimu, Qolla, Lupaka, and Inka. These cultures have left us some traces of their passage on this land. Of course, some of these cultures were more important than the others and therefore, were considered as Mother cultures. In the Andes I am talking mainly about the Chavin in the North and the Tiwanaku in the South.

Through all this time period, the Andean people were united by one common spirit; always honoring the Earth as their Mother. This is why we have found in many of their devices a symbolic and ideological continuity transcending time and space. These seeds, which germinated in various cultures of the Andes, gave their fruits so the Children of the Sun could harvest them, and in their turn, they planted new seeds in order to form the great society of the Tawaintisuyu. Upon the arrival of the new god of the white man, all of this was cut short, ending this time period.

Andean geography is completely different from that of the Old World, which is mostly flat. Here in the Andes, people often perceived the limits of the horizon in the shape of a mountain peak, of a deep valley, of the Amazon forest's green coat, the desert's sands or the sea's deep blue colour. These different perspectives made the Andean people's evolution holistic rather than linear reflecting the whole rhythmic aspect of their surrounding ecosystem. The presence of the mountains maintained the light that shone within them, thus turning them into Children of the Sun and turning their nation into a Solar Culture.

The Children of the Sun's solar culture taught us that "the Andean masters represent the highest dignity, that their thought is people's feeling and their life is the balanced path which allows them to understand nature and the social group as a great community of free beings. To be Andean is to respect all beings who live on the Earth and to consider the planet as a Mother."

We are now in a new Pachakuti, a time of transition and great changes. Our generation and that of our children will be the instruments for the creation of this new world, respecting each living entity, raising the banner of the rainbow, so as to become the conscious life example for our human brothers and sisters.

...We the native masters of the Andes in particular and of the Americas in general, are an alternative for life because of our respect for the dynamic equilibrium and our consciousness of the eminent danger which is awaiting us in ambush, which humanity has not yet known. (Reduced text, Tupaq Katari Indian Movement).

This new Pachakuti is characterized by the presence of the Mother or the manifestation of the planet's feminine aspect, after 500 years of rest. This doesn't mean that women will dominate the world, but rather that men will become progressively more conscious of the necessity to recognize and develop the feminine part of their personalities and hearts. It is only through this feeling of love and balance that Peace on Earth will reign.

This era of Light which has already begun, is emphasizing harmonious life between man and woman. It's obvious that we'll always be somewhat different, but we'll try as much as possible to minimize these differences. Man has to open up to this new planetary vibration of the Mother and feel that our Earth's pole, our center of feminine magnetic radiation, is active and that it is manifesting itself through the doors of the sacred Lake Titikaka.

Since the Mother's energy is active, this means that it's touching the whole world, and those who live on this land and those who come to visit. This is why, sooner or later, when walking on this sacred land of the Andes, we feel the voice of Pachamama speaking within us.

4. PACHAKUTI, COSMIC CYCLE

**Everything comes in its own time,
time and time,
one time, two times, three times
and so for eternity.
(Master White Feather).**

According to Andean societies, life was measured in cosmic cycles, which were called Inti or Sun and lasted a thousand years. An Inti was composed of two halves, each of which was known as a Pachakuti or half an Inti, which corresponded allegorically to half a day within a cycle, identified as a night or a day (a period of obscurity and another of light).

The Pachakuti is the dividing time period within a thousand years, but it also represents five hundred years. During the time of the Inka dynasty, there reigned a king whose first name was Yupanki. He was the ninth king and came to change his kingdom. As he became conscious of the great importance and transcendental aspect of his mission, he changed his name, created the manifestation of the Pachakuti's force and turned this power into his title. In his time, there were many deaths, famine, drought, pestilence and destruction. During several years, the masters did not announce good things. There was much grief and many burials. But he came to change the world turning himself into the Pachakuti, which means "The one who transforms the Earth."

Pacha: Totality, world (space), time, Earth.
Kuti: Revolution, reversing process, transformation, return.
Pachakuti: The one who transforms the Earth.

The king, Pachakuti, came to be and turned the world round. He turned things around, renewing the transformation of the Earth as well as of time and space. Pachakuti is the chaos transforming the world, thus signalling the beginning of a new era, of rest or activity. For the Andean world, a new era started in 1992. During the June solstice of that year, the new Children of the Sun joined to celebrate the beginning of a new cycle, which will take us, on an ascending scale, to the most astounding revolution of our history.

At the end of the past millennium (this book was published in 1997), the new Pachakuti began for the Andean people of 500 years of splendor and glory, a time in which we will be in harmony. The era of the previous Pachakuti ended when the white men invaded this land in 1492, thus initiating a downfall to a period of obscurity, lasting 500 years; in other words, until 1992. It was in that year that Wiraqocha touched the hearts of the new Children of the Sun. It also was in that year that the initiates or *hamaut'as* decided it was time to return and share the ancestral part of their culture with the Children of Pachamama.

The time of power and glory of that previous Pachakuti of the XV century of the modern time, gave birth to one of the greatest sages known in our history. We call him Pachakutiq, the mythical human who incarnated the power of Pachakuti. This was not a simple man, neither was he, only a military chief and strategist and reorganizer of the kingdom. He was also a master, the initiate in the most secret arts of the ancient knowledge of the Children of the Sun, and the visionary of this new era who carved in stone the testimony of his people's destiny. His message has been resting through the centuries, until the moment Wiraqocha chose, at the beginning of this new time, to let his light germinate once again in the hearts and minds of those who were ready to receive it.

Our history is being reborn. There should be no doubt that each and every one of you has been called to add a grain of sand in the growing process of this great mountain that leads to eternity.

The new cycle for Solar Initiations.

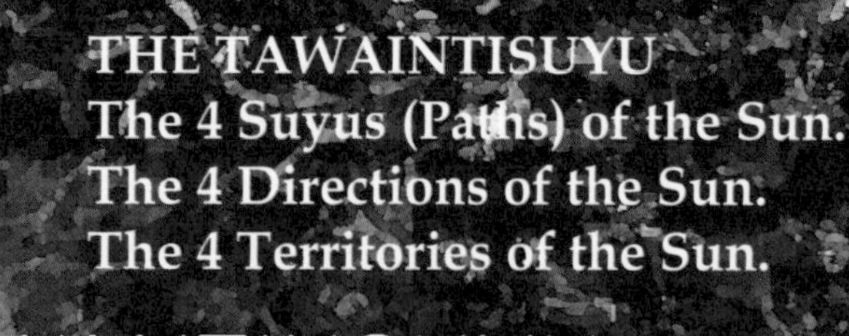

The TAWA-INTI-SUYU (the four sides of the Sun), in the Inka time. Many dispersed ethnic groups from Ecuador, Colombia, Bolivia, Chile and Argentina had been integrated into the Inka kingdom.

5. THE TAWA-INTI-SUYU, LAND OF WIRAQOCHA AND PACHAMAMA

The life of the Universe is as an unending spider web of interconnections. In this microcosm, which is our world, ancient people lived in unity with their surroundings, incorporating in their experience direct contact not only with nature, its valleys, rivers and mountains, but also the dynamic energy of the animal world.

When Western men came to the Inkas' sacred land, they did not understand that life was a continuation of nature's other realms, among them the animal realm, one of the closest to man. The Andean people always sought a harmonious way of life with nature. They respected it, honored it and in this way, included the animals, which they perceived as forms of different vibratory levels. The animal's presence was integrated and assimilated, not only in its ideological state, but also in a practical way. So it was with great pain and remorse that the Andean wise people –the Hamaut'as of the Sun saw how Europeans and clergy understood little or nothing about their respect for the land. They mistook many concepts and identified the Inka culture as a primitive society of polytheist savages who adored idols and probably did not possess a soul, nor the right to be called sons of God. But, the truth always is revealed thus allowing us to reaffirm that our Andean ancestors had an ancient culture and wisdom, not barbaric by any stretch of the imagination. This was revealed by samples of art in ceramic, metal, stone, weaving, etc., which continue to call us with their silent voices.

Wiraqocha was praised in the spirit, manifesting itself to humanity via its son Inti -the Sun. The contact between Inti and Pachamama generated life. Inti illuminated the Andes in the four directions and we saw the sacred Kondor flying over the rainbow. It was then that the Son of the Sun became one with Wiraqocha and Pachamama.

THE TAWANTINSUYU

At the end of the Pachakuti when in Europe they believed the planet was flat, the visionaries of the kingdom of the Sun expanded the name of Wiraqocha to the four edges of the land including one common language for all people. They left their eternal voice in all the stones to form a message for the generations to come. The main expansion of the Inka nation began during the reign of the King Pachakuti, the master who conceived of the Andes as an integrated whole, containing four parts, which he called *suyus*. These were the *Suyus* of the Sun.

*Tawa: Four, the four directions, four elements.
*Inti: Sun, Star of life. King star.
*Suyu: Land, edge.
Tawaintisuyu: The four edges of the Sun. The four lands of the Sun.

WIRAQOCHA

Creator, was at the same time a civilizing entity and light giver to the world. When Wiraqocha created humans, ordered them to go and settle in various places, mountains, lakes, rivers, caves, etc., to turn their place of origin into sacred centers or Paqarinas. Wiraqocha represents the Universe with all its manifestations. Unfortunately, these manifestations were mistakenly interpreted by the church who attributed a polytheist conception of the divinity to Andean spirituality. Wiraqocha is the invisible Universal Being. For the Andean people its physical manifestation and symbol was Inti, the Sun.

*Wira: Untu, the animal fat, the vital energy of the being. Foam, that which is on top of the waves. By extension, that which is white, transparent.
*Qocha: Lake, lagoon, the sea.
Wiraqocha: The foam of the sea. That which is on top of the water, the sea, or Paqarina. That which is white and comes from the sea, is above the origins, the Vital energy of the sea. Wiraqocha is beyond human mental concepts, it's the essence of existence itself.

MOTHER EARTH, PACHAMAMA

Mother Earth gives life to human being, she is the sublime divinity in our world. Pachamama taught us to love everything unconditionally. She showed us that creativity was one of the highest virtues, because if we build with love in our existence, then we become masters.

Pachamama gave us these teachings of life to help us grow. She gave us Munay or Love, Llank'ay or Work and Yachay or Knowledge. Humanity does not need other laws or commandments, because love makes us conscious of Service, which should always be the service of the Being, since the sense of Service is the consciousness of reciprocity or work... And you can be sure that Love (Munay) and Work (Llank'ay), will lead us to the superior consciousness of Knowledge (Yachay).

The invaders who entered on our land created for us laws and precepts of life which were totally in contradiction with the highest principles of community and respect for the land in which the Andean people lived. They introduced into our history the following three laws: Ama Sua, Ama Llulla, Ama Quella - Don't be a thief, don't be a liar and don't be lazy.

A law is created so that a social group can eradicate something bad, so that a population's vices and faults can be controlled.

If people are constantly quarrelling with its brothers and neighbours, it should be taught to love them as if they were their own god! If people live in slavery, eternal conflict, fear and condemnation, it should be taught not to kill and not to wish harm to strangers... I could go on enumerating the laws that are necessary for a society in which people live eternally in an individualistic and conflictive way and whose main instability starts with rejection of the natural world and little respect for the Earth which gave them birth.

What reason could we have to teach a highly developed society, such as the Andean people, not to steal, if that society knows that everything belongs to Pachamama. That society is living by the principle of community life, which is expressed, "All for one and one for all." That society lives by the principle of Service, "Today for you, tomorrow for me." It is not necessary to rob in a society in which the community principle establishes that the children who born will have their own working land? It is not necessary to rob in a society in which the very idea of dying of hunger was neither possible nor permissible, since the distribution of food had reached a highly developed level of organization? What necessity is there to steal?

Why would a highly developed society have to lie and mistrust? In villages where most houses didn't even have a door, much less security systems, why would there be the need to lie? Why, if that action does not help us grow spiritually?

Also, how could we say: "Don't be lazy?" to a society which built all its grandeur, stone upon stone, to that which knows it is only through its work that it will survive and be great?

The Christians of the past had to somehow legitimize all the thieves, lies and laziness, they were responsible for on this land.

We also know that Pachamama is Mother of all purification, cleansing and pardon. We have now begun a new era, and in this era of light, all brothers and sisters are welcome. Let's allow Wiraqocha to touch our Inner Sun so that through the emergence of conscious love within, we may be men and women of this new era.

PART II

THE SERPENT, THE KONDOR AND THE PUMA

The Snake master, an anthropomorphic being (Chavin).

A Puma or Jaguar with square crosses (Chavin).

1. CHAVIN OF WANTAR

Considered one of the Mother cultures of the last two thousand years, Chavin culture originated in the mountains of Northern Peru, with the influence of this great culture reaching Northern and Southern Peru to the shores of the Sacred Lake Titikaka, saluting the great Wiraqocha.

In addition, the Tiwanaku culture preserved this evolutionary line and cast its seeds in the Cusco valleys to preserve, with the Inkas, the silence in their voice, a voice that emanates eternally through the centuries.

The people of Chavin preserved an ancient heritage. They learned that within Pachamama there was an underground force that was moved by the Great Amaru or Serpent. They knew that one day this Amaru would open its wings of liberty and ascend in an undulating and spiralling path up to the house of the Amaru Tupaq or the level of the royal light.

Working with the Amaru, it was necessary to feel its movement and see its path, because this path would lead the children of the highest Apu (mountain spirit) in the Peruvian Andes, the Waskaran (6768 m), to integrate the Puma into their spiritual practice, as the entity of stability. In turn then, the Puma would project people's leap to superior planes, where the Mother Bird reigns.

The symbol and power animal used by the Chavin people was the Amaru or Serpent. The Serpent was an important part of their culture and, in their art, we find this Serpentine symbol repeatedly in ceramics, stone work and other graphic arts. The Serpent is identified with the first evolutionary stage in humanity. It ignites the force that slowly ascends from the coccyx, up the spine, to the top of our head, to our crown where ecstasy and higher consciousness can be achieved.

The Chavin masters also knew about the feline movement and its metamorphosis. Someday, the feline would reach the highest summits of the Apus and from there would salute with wings of light, the Sacred Bird of the Sun, the Kondor.

Wiraqocha, symbol of the Solar Being (Tiwanaku).

2. TIWANAKU

The Tiwanaku culture was born in the immediate surroundings of Lake Titikaka, with the energy of the Achachillas -mountain spirits- in the Royal Cordillera. As a civilization of the high plains, it knew well the movement of Katari or Serpent. They used the Earth's magnetic force and gave their salute to the planets, stars and the Sun. Kantataita or the Sun's star accompanied humanity to their destiny, until the day the Sacred Mallku -Kuntur Apuchin or Master Kondor- showed them the way to Ausanqati, the main Apu in Cusco territory.

Tiwanaku represents one of the Mother cultures in Southern Peru. Lake Titikaka nurtured this society and their culture, sending them to other lands with the mission to expand the people's consciousness of Wiraqocha and its Solar science. Wiraqocha is the being of polarity who controls that which is above as well as that which is below, what is on the right and the left, that which is masculine and that which is feminine. Wiraqocha is the all-inclusive entity that represents time and space.

In Tiwanaku, they represented Wiraqocha as a Solar Being, who, as well as having the fiery presence of the Serpent, manifests the stability of the Feline Totem, the Puma. But it was the Kondor that was immortalized as the Sacred Bird.

Tiwanaku´s iconography is in some parts similar to that of Chavin, as well as of other great Andean cultures. This allows us to see the uninterrupted evolutionary continuity in many aspects of its spiritual and ideological development.

The most prominent animal in the ideological and spiritual concepts of the Tiwanaku culture, was the Sacred Mallku or Royal Kondor. The Kondor is without doubt (along with the Q'ente or Hummingbird), the bird that figures most frequently in Andean cultures; but it had a very special meaning for the people of Tiwanaku. The Kondor is considered the bird of the Sun that sees all and visits the most distant spaces of our terrestrial sky, until it loses itself and joins in that great unity called Wiraqocha.

Paka Haqe (Eagle Being).

The most ancient salutes of the children of Wiraqocha. The right hand is placed on the heart center and the left hand is resting on the solar plexus.

Mallku Haqe (Kondor Being).

Sunrise, Sacred Lake Titikaka.

Inka wall.

3. THE INKAS

Inti: Sun, sunray.
Khana (Regional Aimara): Light.
Inkha (INKA): Solar light.
Children of the Sun.
Children of the Light.
Son of the Sun.

Known as the Children of the Sun, the Inkas belonged to the solar hierarchy and integrated the Andean cultures. They united the most distant societies within the four corners of the kingdom and made one of the largest nations in the world.

The Inka culture, as an organized society, was, as far as we know, perhaps the most advanced of all its predecessors. Of course, they were not perfect, otherwise they would still be present and determining the destiny of our society. In their valleys and mountains they built in the hard stone, showing their respect for the land and full integration with their geographical surroundings. In this understanding of its ecological vitality; in a practical way they applied the ecological principle, which is harmonious life with all that surrounds us. This is the practice referred to as Munay, Llank'ay and Yachay. Their understanding of Llank'ay, or work, was integrated in each action by chanting and praising, saluting the Sky and the Earth. What could we really teach of ecology to the Andean ancestors, if they were themselves born ecologists?

Wiraqocha came to Earth, and from the foam of the waves of Sacred Lake Titikaka roused the mythical couple, Mallku (Manqo) Qhapaq and Mama Oqllo, sending them to search for the favorable land which later would shelter the Children of the Sun. They followed Wiraqocha's path in a pilgrimage that took them from the South to the North, thus reaffirming the path of initiation known as Wiraqocha's route. This mythical couple, Mallku Qhapaq and Mama Oqllo, walked in Wiraqocha's presence accompanied by the Sacred Kondor and arrived in Cusco's valleys, where they began to create one of the greatest kingdoms on the planet.

When Wiraqocha's children arrived in Qosqo (Cusco), they set foot on the Puma's territory; the Puma that would help them transcend to the most unimaginable planes.

The evolutionary cycles show us that if a culture rises today, tomorrow it will begin its downfall. This downfall does not mean there is regression within the being, but rather a braking point; and as far as cosmic time is concerned, no more than an instant. Sometimes, it is necessary to stop awhile, in order to evaluate the progress already made. These evolutionary cycles in the Andean world were known as Pachakuti.

It was at the beginning of the Pachakuti when in Europe were in the darkest time of their history (XI century of modern time) that the slow but decisive birth process of the Inkas began to take shape, Mamaqocha (Lake Titikaka) was preparing her lap, so that in Pachamama's womb, there would be important Paqarinas (places of birth) for these Children of the Sun.

During these evolutionary cycles, the sacred totem animals also changed with the new Pachakuti. Each evolutionary cycle was marked by the presence of an archetypal animal; serving as the most important symbol within the spiritual development of the society.

With respect to the concept of archetypes within a society, it should be clear that, for example, in the Chavin culture if the masters were decorated with snakes, this did not mean that they ate insects or crawled on their bellies. Rather, the archetype of the Amaru or Serpent indicated the type of development that the individual was expected to undergo. From within himself, from the Ukhu Pacha (Inner World), with a movement that is controlled and secure he would hold his head erect, proudly, as the serpent does.

The Tiwanaku culture projected itself towards the sky, with the help of the Sunbird, the Kondor. This archetype guided them to the infinity of the space surrounding them, and so they made friends with Kantati Ururu (Venus), with Wara Wara Ichawa and Chaka Silltu (constellation of Orion and its stars: Delta, Epsilon and Sigma), with Kaura Nayra (eyes of the Llama), with the Chakana (Southern Cross) and other stars and constellations. The Kondor also taught them to see through space and to recognize their inner stars. The Kondor's archetype had overtaken the Serpent and in its path, gave it wings to include it in the world of Hanan Pacha. The Kondor is the bird that always went with the Achachillas (Spirits of the mountains), and occasionally came down to Hatun Qocha (Great Lake). In an ascending return flight, it was transported with the help of Wayra -the Wind, to the highest peaks of the Andes.

The Children of the Sun arrived in the Puma's land of Cusco, at the beginning of the Pachakuti of the XI century of the modern time. The Inkas followed the Amaru's path, but when they saw the Kondor's flight, they fell in ecstasy with that tremendous spectacle keeping them in an ecstatic state. From then on, they slowly developed the consciousness whose flowering became The Awakening of the Puma.

Sunrise (Lake Titikaka).

PART III

THE AWAKENING OF THE PUMA
INKA INITIATION PATH

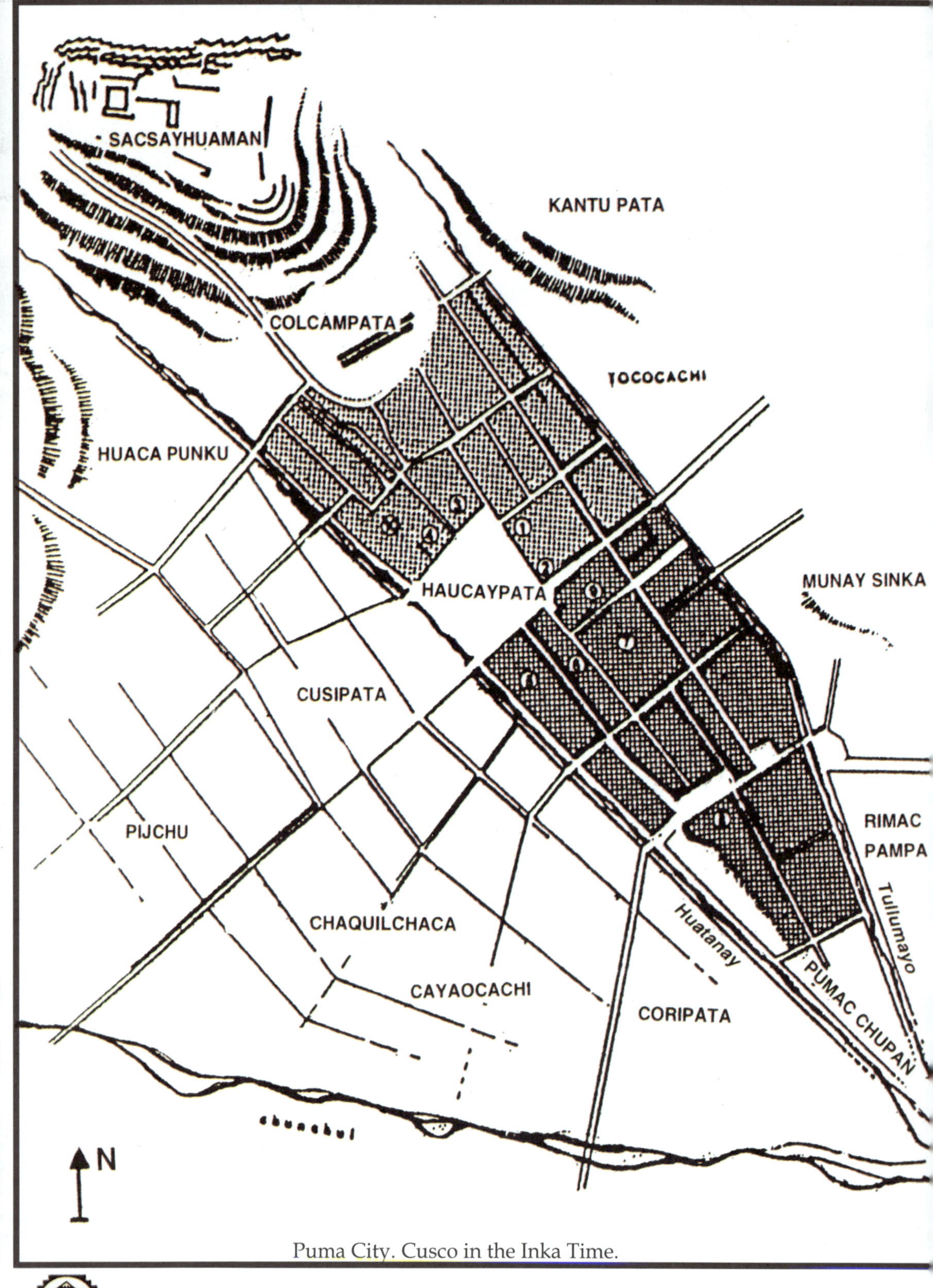

Puma City. Cusco in the Inka Time.

Inka wall in Cusco.

1. CUSCO, PUMA CITY

Designing a city where space distribution is identified with the organs of an animal, and whose overall shape is that of the animal in motion, required the application of multiple sciences at their highest level of conceptualization and practice.

The construction of the city of Cusco represented the synthesis of a great period, the golden years of the Inka culture.

Its urban beauty was an expression of a combination of art and nature; its construction integrated itself with the city's Northern hill, the House of Saqsaywaman, defining the head, the upper part of the totem animal. The city of Cusco was magnificent from all directions. Seeing it, the Christians of the past were fascinated as well as awed to find a seemingly primitive village that could easily compared to any European capital of the time. The silent words of the Inkas, in each stone wall, indicated that they were far from savages. On the contrary, Cusco is a city that neither the Christians who invaded it nor modern societies could understand.

Pachakuti was the architect of the sacred city of Cusco, was the ninth king of the Inka dynasty. He was not only an architect, military strategist and leader of his kingdom, he was also much more, so much more as an initiate in the master path! The initiate is the one who lives in harmony with the universal essence.

Pachakuti, the great master and transformer of the kingdom of the Children of the Sun, came to reign before the end of an Andean cycle. When he came to power, he announced the advent of a New Pachakuti, and of course the end of the previous cycle was marked by the arrival of the invaders to the "New World known later as America." The coming of this great master was supposed to close a cycle and generate a new one. This new cycle would be marked by the expansion of the Inkan kingdom, reinforcing its walls and buildings, such as in the city of Cusco, during Pachakuti's government.

A genius and visionary, Pachakuti changed his father Wiraqocha's simple city into the nation's mother city, the heart of the Andes. The ancient city became the Puma City. The Puma became not just a symbol of power and expansion, but the principal archetype of the Inkan culture, more important than any other animal archetype with the exception of the Kondor, Serpent, Llama and Hummingbird. The Puma also represented the image that the kingdom of the Children of the Sun took to enlarge and enrich its territory, and thus form the Tawaintisuyu.

The Puma city became the richest and most important in the nation and gave honor to its reigning Star when the Sun touched it each morning at dawn to start a new day.

Because Pachakuti -the great master and Son of the Sun and the stars, knew their movements and manifestations, when he designed the Puma city, he planned its orientation with an eye to the path of the Sun and Wiraqocha's stars.

Inka building (Qorikancha Temple).

Inka wall (Qorikancha Temple)

CUSCO, THE SOLAR PUMA

Cusco was the Puma city, and the Puma as an entity of power and archetype filled every inhabitant and visitor with its energy as they set foot on its grounds. The people of the city, and progressively the whole kingdom, had to vibrate in unison with the feline totem. At some moment in their lives, they would then be blessed with the Puma incarnation.

Every 20th of June, the Pumarunas (Puma men) of the sacred city of Cusco, prepared themselves to honor the rising of the Sun on the following day. The great Pachakuti had laid out each part of the city so that all of its temples and palaces would receive the Sun's light in a predictable and progressive pattern.

It's true that the solar science, in its more complex stages, belonged to the Hamaut'as and Yachaqs (the masters). On the other hand, ceremonies such as the Awakening of the Puma City of Cusco, showed that the inhabitants had common knowledge of the path of the Sun. Even the children participated in celebrating the new solar year each June 21st when the Sun began its journey away from its Northern most rising point, tracing its way to the South.

Pachakuti, as he was designing and directing the construction of the Puma City, left us a message for the time of the New Pachakutis. During this new time period the great Solar Puma awakens each year, and when the qosqoruna (the inhabitant from Cusco city) becomes conscious of this event, he will turn himself into a Pumaruna, thus recovering his stability and transcendental consciousness. The Puma will give this new man stability and harmony with his surroundings. The Puma will give him the intuition of what is transcendental. The expansion of his consciousness will allow him to be Amaru (Serpent), Kuntur (Kondor) and Puma at the same time.

The Ukhu Pacha (Inner World of the Serpent) and the Hanan Pacha (world on the Kuntur) will be experienced in the here and now of the Kay Pacha (world of the Puma). The Earth plane of Kay Pacha belongs to our dimension and it is in that dimension that we must integrate the Ukhu Pacha and the Hanan Pacha.

The trinity concept as an ideological element does not belong exclusively to any culture of the world, but rather to all of them. With each step humanity takes in life, feel confronted with spontaneous and natural manifestations of the number three.

It's logical that in the ideological interpretation of life and the world, there were some cultures that were more advanced than others. Some of them defined their cosmology in a rational and logical way, by man's immediate reality. In other words, our life's continuity is only possible when there is a harmonious encounter between a Man and a Woman, producing a Child. When a masculine charge mixes with a feminine charge, this creates a unity and the result is continuity; in this case, the child. Apart from that, only a few cultures believed in the mixing of two positive charges, such as the Father and the Holy Spirit, to create the third member of their trinity, the son. It's a shame that these unbalanced concepts of nature still exist and enjoy popular acceptance today.

The Andean culture primarily respected the Mother, and so it defined the three energy levels of our Universe with the following names: Ukhu Pacha, Kay Pacha, and Hanan Pacha. The meaning and important details about these planes will be discussed in the book "Machu Pijchu Forever."

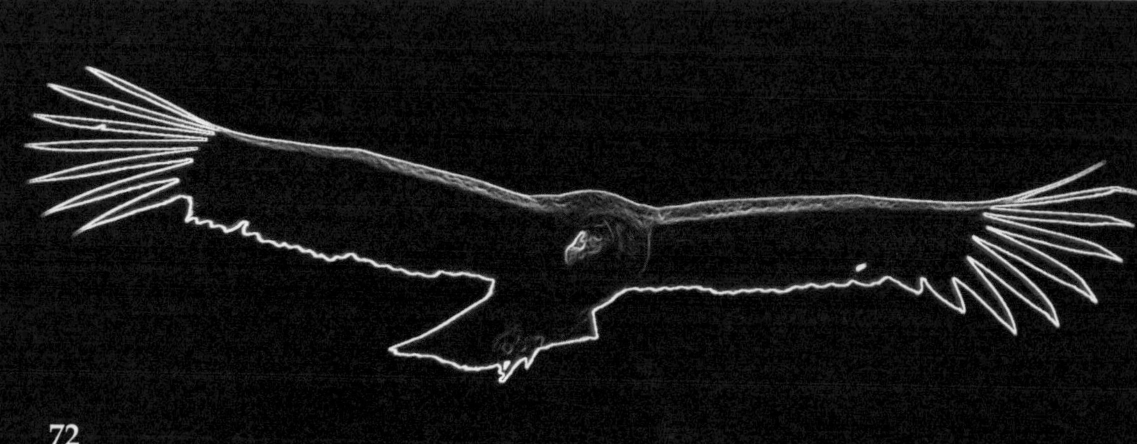

the Earth, in its planetary movement, had reached the farthest point from its center -the Sun, and was about to start its return, repeating the cycle and initiating the new solar year.

On this great day of June 21st, the mountains not only were the community's guiding Apus (spirits of the mountains), but they also participated in the sacred dance to Inti. From near and far, the people honored the Apu Pikol (mountain spirit Pikol), which on this day had the privilege of showing the way to the Sun. Meanwhile, the Apus Pukamoqo and Socorro, his pages, would salute the Children of the Andes, and soon helping the Hatun Puma to completely awaken.

The geographical position and the wavy shape of the guide Apus Pukamoqo and Socorro, in relation to the Puma City, would serve as guides for the rays of the Sun as it shone down on the sacred city.

From the head of the Puma-Apu Saqsaywaman, the sounds of the *pututus* or ceremonial chants could be heard. In the distant folds of the mountains and in the sacred city, the pilgrims who had come from all over the kingdom and taken their positions, were absorbed in a deep meditation. The Hatun Puma or Cusco city was slowly and harmoniously awakening. From the higher parts of the hills and from the head in Saqsaywaman, they could all rejoice in observing how the shadows on the city slowly disappeared. At a certain point in its progression, the Sun would touch the Puma's tail (Pumaqchupan). At that moment, contact was established between the upper part of the Puma and its lower extremity. The masters knew that in this way, the Sun would activate the Power Entity, the Puma, and the Sun's light would ascend from the bottom of the spine in the tail, to finally meet at the crown of the head which was already lit.

If groups of people are positioned along the Pumakurku street (the spine of the feline, the column through which the Amaru's energy ascends) they will be part of this worthy manifestation of the Sun, which, after touching each street oriented to the East moves on to awaken the Great Puma. This manifestation of the Sun is astounding, as it slowly ascends through the spine of the city, with the head in Saqsaywaman already illuminated.

A few years ago, as I was carrying out these studies, I was sad to see that it was no longer possible to investigate many of these phenomena. My face was marked with nostalgia, because my blood brothers had helped the invaders in their systematic destruction of many palaces and temples, without any respect for the cultural and scientific development of our people. What type of people were they that did not even care to take the technical and scientific development of the culture they invaded? Fortunately, they didn't destroy everything and from the shadows of the night, the walls built by our ancestors still stand, as sentinels awaiting a new dawn.

I would like to show that as part of the systematic destruction of our temples and *wakas* by the invaders, they also tried to generate an atmosphere encouraging the mental degradation of our pueblo. The church condemned our sacred power places, calling them centers of Satanism. Our cultural centers were presented as ridiculous and unimportant; our astronomical observatories were said to have questionable, vile functions. An example of this degradation is the mistaken title Horca del inca (gallows of the Inca) given to the astronomical observatory in Copacabana. This sacred power place is one of the most amazing Intiwatanas (solar observatories) in the region of Lake Titikaka, indeed in the Andes.

In this sacred place the path of the Sun appears on June solstice every year. At this Intiwatana, the Sun shines through a small hole, creating the first astronomical sign on an altar made especially to observe this phenomena. Solar observations in the Titikaka region will be discussed in more detail in upcoming books.

I will show photographs and drawings of the astronomical event that will illustrate the effect of the interesting phenomena I call the Awakening of the Hatun Puma.

Beautiful sunrise in Cusco.

Serpents with
Puma-Priest
(Mochika).

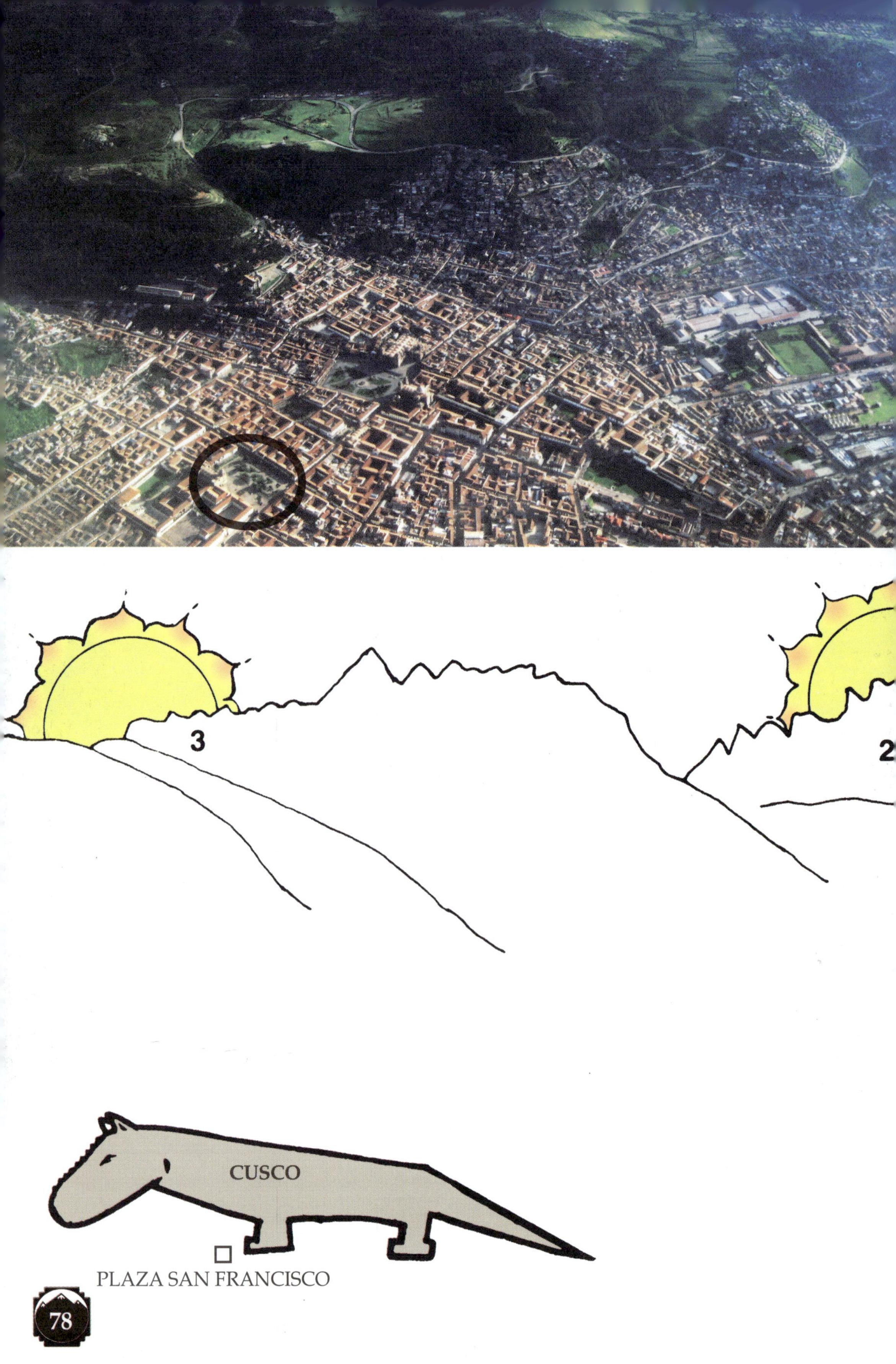
3
2
CUSCO
PLAZA SAN FRANCISCO

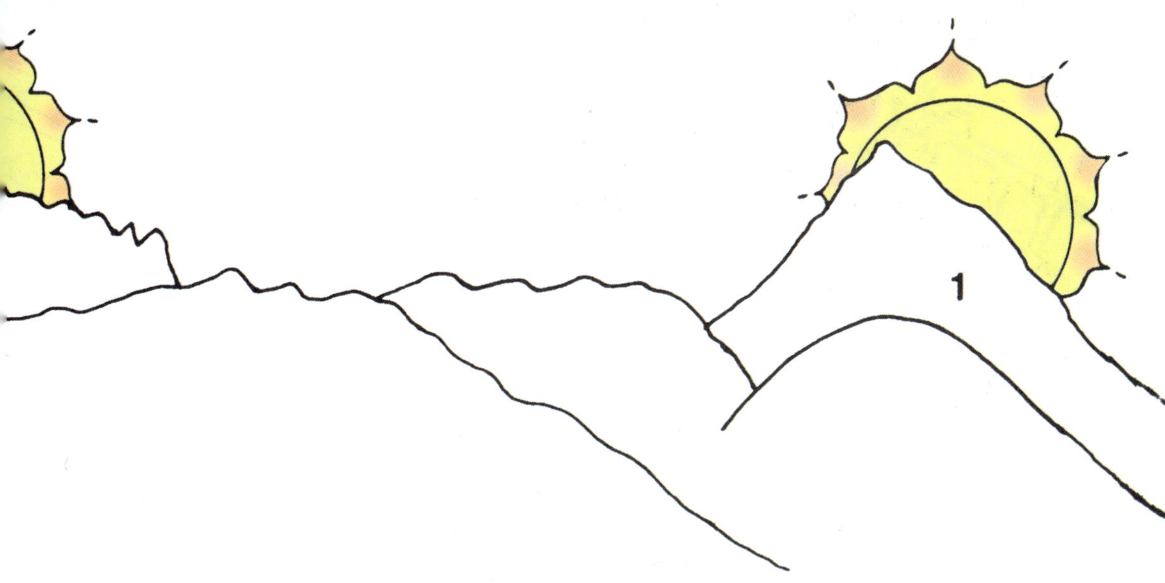

Reference points of Sunrise at different times of the year:

1. In December solstice, at the right of the Apu Ausanqati.
2. In September and March equinoxes, behind the Apu Pachatusan.
3. In June solstice, at the left of the Apu Pikol.

* Reference point: Plaza San Francisco.

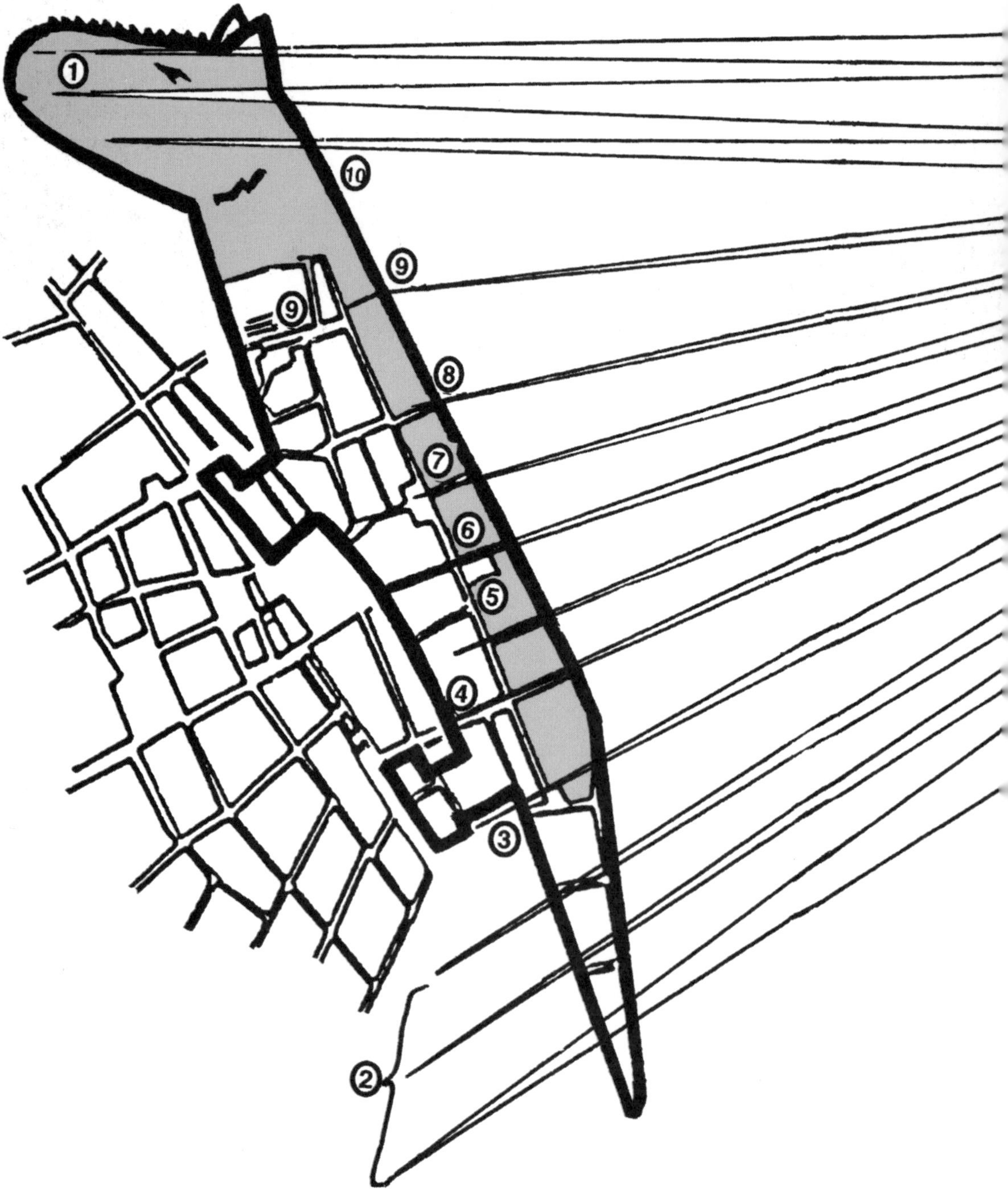

Process of illumination of the Hatun Puma during June solstice.

1. The head is illuminated (Saqsaywaman).
2. The tail is illuminated (Pumaqchupan).
3. The walls of the Qorikancha indicate the moment of Sunrise.
4. Maruri Street receives the Sun (Tupaq Yupanki's palace).
5. Inka Roq'a Street indicates the direction of the Sunrise.
6. Hatunrumiyuq Street is illuminated (Inka Roq'a's palace).
7. Siete Culebras Street is awakening; indicating the Sunrise.
8. Ladrillos Street is aligned with the Sunrise.
9. The aligned walls of Qolqanpata indicates the direction of the Sun.
10. The Sun meets again with the head of the Puma awakening.

* Reference point: Follow-up along Pumakurku street (The Puma's Vertebral Spine).

Illumination of Maruri Street (Pukamarka palace walls). June 21st.

Illumination of Inka Roq'a Passage (Inka Roq'a's palace). June 21st.

Streets already illuminated along Pumakurku (South). June 21st.

Illuminated Siete Culebras Passage; Pumakurku to the North is still without Sun. June 21st.

Sunrise on June 21st in Siete Culebras Passage. All these solar effects marked the coming of June solstice and the beginning of the Inti Raymi -Festival of the Sun (pages 84 and 85.) >>>>>

Waka of Q'enqo.

Q'ENQO - SOLAR OBSERVATORY, THE PUMA'S WAKA

It is obvious that the information given by writers and historians concerning many archeological centers throughout the Andes was at fault or incomplete. In the particular case of Q'enqo, I'll shed some light on the subject, based on the observations, measurements and studies which have been made during the 80's and 90's. Let's look at these magnificent solar events. Writers referred to solar observations made by the Andean people. Historians and other experts agree on this point, but during the last five hundred years of cultural and physical domination, information could not be released, neither could all sciences be involved in field studies.

Certainly what I'm presenting here is not in relation to the whole site of Q'enqo, but does correspond to its astronomical function at the level that I discovered since 1993, there are surprising phenomena of light and shadows. In the upper part of the massive block of limestone, there are a couple of *sukankas, saywas,* gnomons or a sort of Intiwatana (gnomons, the raised part of a sundial that casts the shadow). This is our primary area of astronomical study. June 21st is the most important date for solar observations and astronomical studies here in the Andes. There are, two dates which could hold the honor: June solstice and December solstice.

December solstice was the beginning of the great ceremony and festival of Qhapaq Raymi. That day marked the start of a new season, a season that had much to do with crops germinating in the ground.

December solstice was a good reason for celebration. One season later, the first harvest of the seeds would take place; these fruits would continue to feed the Children of the Sun. So, on that celebrated day, a life cycle began which was important for the people of the mountains, valleys and rainforest. These people's lives would be accompanied by the presence of water with the arrival of the rains. On the day of Qhapaq Raymi, the terrestrial new year started. This was the time when the first great rains renewed the earth, and little by little, the clouds hid the mountains, thus preventing the Sun's rays from reaching the Temples and *Wakas* which were lined-up along the different *seq'es*. This was not the best time for solar observations, because of the differences in visibility during the rainy season in December.

The Inti Raymi took place on June 21st (solstice). In a clear blue sky, the Sun followed its course and with each second, the sacred sites were touched and brought to life. The continuous movement of this solar cycle manifested through contrasts of light and shadows, creating angles, measures and forms which all had meaning. The best season for solar observations is when the voice of the rain remains silent. In the Andean Calendar June 21st was important for many reasons.

Some reflected an earthy level, and others clearly had a heavenly connotation. The planet had reached its farthest point in relation to the Sun, and on that day would start to slowly move closer again, so that it would, in a symbolic way, remain attached to the Sun and thus would always accompany Inti -the Sun.

Andeans, like many other ancient people, built sites that were used for observations and measurements of time. Many cultures, in other parts of the world, also created specific places to follow the movements of the Sun and other planets in the sky. Some of these constructions were very simple, but others were so complex that they could only be the work of wise people or masters.

There were ancient cultures like the Egyptian, Chinese, Hindu, Maya, Aztec, Greek, Sufi and others that were lost in time that distinguished themselves by their extensive knowledge of astronomy. For instance the Maya astronomers built pyramids in order to mark the arrival of the most important days of the year. Today on the Yucatan peninsula, you can still observe how at the equinox one side of the Pyramide of Chichen Itza is illuminated, creating the shape of a gorgeous Serpent that is seen as coming from the sky to encounter the head that was sculpted on the very bottom of that side of the pyramid. This solar phenomenon is not just astronomical but a very important visual evidence of the Mayan cosmology.

In ancient cultures, the most practical way to learn the basic principles of astronomy was to put a wooden stick in the ground or in a flat area to form a 90° angle with the earth. This stick, or in other cases, a column which was specially built near or inside a temple, would project a variety of shadows as the days went by. Taking one of these projected lines, especially those which are located at the extreme (and are projected on the solstices), we'll find that the Sun, in order to produce another identical shadow, will require approximately 365 days. So the observation of shadows produced by a column, stick or gnomon, is the simplest practice of astronomical science.

The Inkas and other Andean peoples also knew this basic system for solar measurements, but the linear observations were not sufficient for them. They created more sophisticated systems and built specific places for this purpose. With these people, astronomical studies reached a very advanced and yet unsuspected level. Let's take for example the gigantic network of lines in Nasca's pampas, where the engineer Paul Kosok was surprised to discover some of the projections of these lines on the ground as well as in space. Some of these lines are oriented in the direction of important volcanoes and Apus -Spirits of the mountains, others are projected towards the points where the Sun rises and where it sets, others signal the path of a constellation or the Moon. (Nasca and other enigmas will be discussed in another book). Like the Nascas, other Andean peoples, all Children of the Sun, had to know about the Sun's manifestations. In Tiwanaku, the astronomical knowledge was admirable, and today, the villagers of the Altiplano (high plains), still come together to celebrate important days, such as June 21st to remember their past and be touched by the Sun, as a symbol of this cultural continuity.

Q'ENQO, THE AWAKENING OF THE PUMA

In order to prove that the Inkas had elaborated sophisticated systems to conduct their astronomical observations, I must first demonstrate that in some of these solar revelations the evidence in light and shadow is not limited to simple linear projections. Rather, the line transforms into definite shapes representing living beings. These in turn, go through a metamorphosis to produce man's enlightenment as an ultimate result.

Q'enqo will be one of the paths travelled in this world of shapes, showing us with solar evidence, that at some moment in their history, our ancestors showed us the way to transcendence.

On June 21st (Inti Raymi), Q'enqo's *sukankas, saywas* or Intiwatana serve as time meters. The manifestation of light and shadows will go through constant changes up to the point of integrating human being with his surroundings. The Sun's most Northern position, will make it possible for this *waka's* sleeping feline to awaken. The small stone wall oriented to the Northeast, in relation to these *sukankas,* will be the key site for these effects of light and shadow. It's on that small wall which is handworked and naturally shaped, that the first solar manifestations appear.

Sunrise in June solstice (Q'enqo).

Sunrise in June solstice (Q'enqo).

The first manifestation on this Intiwatana takes place because of a narrow fissure in this small wall with the Sun's rays passing through this fissure and touching the *sukanka* on the left (relative to the Sun), with a line of light. So this line, defined by the presence of the Sun, is the first indicator of this astronomical phenomenon. This same line almost at the corner or extreme edge, which was projected on the small platform of the *sukanka* at the left, will progress forward until disappearing from the upper platform and continue to move down to the base that supports the two *sukankas*. When this first line, projected by the Sun through the fissure onto the *sukanka* on the left, continues to move and arrives at the base that is immediately below the *sukankas*, it is as if it is following an important mark. When this line travels and arrives at the middle of the separation of the edges to the right and to the left, the illumination of the upper platform begins again, producing a progressive result that is seen as the representation of the left eye of a power animal. This illumination will grow, and when this line arrives again at the extreme edge, as if touching the border of the middle base, the illumination of the second eye, which is the *sukanka* on the right, will begin. Finally, after the first ray that shined on the upper *sukanka* has passed down to the lower base of this solar instrument on the ground, the two eyes will be entirely illuminated to Awaken the Solar Puma of this great center. When both platforms are entirely illuminated, the feline totem, the Puma, is fully awake and its face is completely defined. As soon as the observer is ready, it will introduce the Awakening of the Master.

If we observe this solar instrument or Intiwatana, we'll see that it was constructed on three levels corresponding to the Andean Cosmic Vision according to the division of the Universe.

*The first point-line made by the Sun at the upper part of the *sukanka* on the left in relation to the Sun, was the first touch of the higher level called Hanan Pacha, so the solar Inkan initiates were controlling step by step the worthy arrival of June solstice. Legends say that the Sun came down to our world to sit on its throne, meaning to take its place at the Hanan Pacha, the Kay Pacha and the Ukhu Pacha. With the first beam of light the Sun was seating itself at its first and highest level, the Hanan Pacha.

*After the illumination to the edge of the upper part of the *sukanka* on the left, the line produced by the fissure travels down until arriving at the middle part which symbolically marks the second level or throne of the Sun, which is the Kay Pacha. All this evolution is important in order to understand the higher secrets of the Andean leaders. Remember that when this line arrives almost at the middle, the illumination of the upper part of the *sukanka* on the left begins; this is the left eye of the Solar Puma.

*This line continues to move, and upon arriving at the left border of the base that supports the two *sukankas*, the illumination of the second *sukanka* on the right with respect to the Sun begins. It is only when this ray, after entering through the fissure arrives at the base on the ground, that this solar instrument once again produces the solar phenomenon of the Awakening of the Puma and the Master. So finally the Sun touches the base that is on the ground as it sits upon its throne and completes this manifestation while marking the third level that corresponds to the Ukhu Pacha.

As the Sun touches the eyes of the Puma, this symbolically means that the feline totem is directly converted into an initiate by Wiraqocha. So The Awakening of the Puma will serve as a model for the awakening of the Superior Human Being. In other words, the Puma transforms itself into a Person; and not just into any one, but rather into the Solar Initiate, thus allowing to each person to regain the condition of being a Pumaruna, one who integrates the three worlds. This Pumaruna is the Hamaut'a, who will appear in the very same *waka*, a few minutes after this great awakening. The presence of the feline in the pre-Columbian cultures is constantly present. We can find it in the majority of devices such as weaving, ceramics, metal and stone work.

If we go for a walk in the streets of the Puma city of Cusco, we'll be able to see for ourselves that along some of these streets there still are felines carved in stones and embossments.

In the Inkas' time, it is certain that the majority of the walls were decorated with visual symbols such as birds, felines, reptiles and even people. If we take a walk in the Andes, we will see that in many archeological sites, a variety of images are still present. Once again, the invaders wanted to destroy our people's entire material and ideological heritage with the capital city suffering the most. They tried to destroy everything that contributed to the development of this culture. The symbolic representations of felines and birds that existed in the city were rich cultural elements that they feared might go against the domination and indoctrination decreed by the religious crown. If it were possible, they would have destroyed the Puma and the Kondor, so as to only leave the Serpent as evidence of a spiritual cult which, in the biblical dogma represents the devil. It was logical that if a culture honored the Serpent, it was associated with Satan himself. This is the reason why, when we walk through the city's streets, the majority of the symbols that are still apparent are the Serpents of the Ukhu Pacha.

The Solar Initiations with phenomenon in this *Waka* of the Puma will be described in another book, I will also introduce some more secrets about this Sundial in other periods of the year such as equinoxes and the December solstice.

Let's see on the following pages the photos and drawings of The Awakening of the Puma or of the Totem Feline which takes place in Q'enqo's *Waka*, on June 21st, day of solstice, the original celebration for Inti Raymi, the Festival of the Sun.

Beginning of the astronomical phenomenon. The Sun enters through a fissure, touching the *sukanka* almost at the edge on the left.

Sukanka on the left is illuminated by a line of Sunlight, this first point will transform into a line located almost at the edge.

Illumination at the left border of the base that supports the two *sukankas*, the illumination of the second *sukanka* on the right with respect to the Sun begins.

The point-line made by the Sun touches the base that is on the ground and produces the Awakening of the Puma.

The Awakening of the Puma and its artistic representation, an exceptional phenomenon which occurs every year on the day of the New Solar Year, June solstice at Q'enqo, *Waka* of the Puma.

Representation of the Master Bird.

Parallel to the Awakening of the Puma, here is the complete illumination of the smaller feline which is located a few meters away to the West in Q'enqo.

Q'enqo, zigzag, channels, also related to the labyrinth of life and represented by the Amaru-Snake.

Serpentine channels in Q'enqo, note the head of the Hummingbird emerging from their converge, also illuminate in June solstice.

Alchemy.

INITIATION PATH
THE SAGE, THE HAMAUT'A AWAKENS

The connotation of awakening is related to the state of Consciousness or Conscience, meaning With Awareness or with magical knowledge. It is very surprising to see that modern civilized man, the so-called knowledgeable, rational man, only considers his life in a scheme of linear development; that is within the limits of what he can see, feel, smell and touch.

Someday, relatively new sciences like archaeo-astronomy, quantum physics, magnetic measuring and others, will be part of modern man's common life. He will no longer be surprised when studying ancient civilizations, because he will have the means to prove that our ancestors used pure science which at times was very sophisticated. Of course, we have the right to question, but doing so, we also have the responsibility to investigate first and to avoid oversimplification and erroneous conclusions.

In the Andean world, the greatest knowledge was entrusted to the Hamaut'as, the sages or masters. In a certain way, the Hamaut'a was the voice of the Sky and of the Earth. He/she was the master who would always walk with awakened consciousness and sufficient energy to help the Son of the Sun in directing his people.

The solar manifestation and its astronomical precision in Q'enqo's Intiwatana is worth mentioning. We have seen how the Sun produces a variety of shapes in a particular space, these shapes are the result of light and shadow. On June 21st, in this Intiwatana, a few minutes after the manifestation of The Awakening of the Puma, we see the Puma change itself into a Sage. The Sage or Hamaut'a will define himself through light and shadow, the climax being is the primary state of Initiation.

In the surrounding area, the sages built a whole series of seats, thrones and steps. Some of these thrones were functional and were used in those moments of increased energy, so that anyone who was touched by Wiraqocha could become a Hatun Runa, or Superior Being, whose guide would be the Hatun Puma or Solar Puma.

The necessary references to understand the magical manifestations of the Awakening of the Sage, are not only available in Q'enqo, but also at additional places such as Ollantaytambo, Machu Pijchu, and others. In Saqsaywaman, in the East near the Chinkana Chica area, there is a type of small amphitheater where they carved seats

The Sage, Initiate or Hamaut'a when Awakes and artistic work representing the astronomical phenomenon. June solstice at *Waka* of the Puma, Q'enqo.

and a series of steps and walls. On one of the sides in a great block of limestone, a human profile has been sculpted, following the natural shape of the massive stone, which has been given various nicknames: The sentinel, the warrior, the Inka. In fact, this megalithic piece that often remains unnoticed, carries a spiritual meaning and on important dates fulfil the purpose for which it was carved. On June 21st, before the illumination of the face and especially at the level of the eye, the upper part of the *maskaypacha* -royal signs- lights-up (See photos).

The Inka-warrior in Saqsaywaman with *maskaypacha* illuminate.

Sunrise during the equinoxes in Q'enqo.

3. THE DIVISION OF THE WORLD SEPTEMBER AND MARCH EQUINOXES

I have already written about the equinoxes in the first part of this book.

Ideologically speaking, in symbolic language, I'll call the equinox the division of the world. Once more we'll be going to Q'enqo, to its Intiwatana (*sukankas* or *saywas*). Intiwatana means the place where the Sun is attached or the place where the Sun makes one year. Symbolically, these usual meanings are correct. First of all, the place where the Sun is attached or where the Sun attaches itself, indicates that the Sun always comes back. More accurately, the Earth is attached to the Sun. In the second meaning, an Intiwatana is a place to which the Sun returns after a round trip of a whole year. This verifies the solar year.

If we consider the solstices as the extreme points of an arc made by the Sun, then, the in between points will mark the equinoxes. The displacement along the horizon from one solstice sunrise to the other is approximately 47°. In the case of the equinoxes, the measure is half way between them because in the Southern Hemisphere on September 23rd (Spring Equinox) as well as on March 21st (Autumn Equinox), day and night will have the same duration as they do in the Northern Hemisphere. So, we'll consider that this phenomenon is objective, and the observer's geographic location relative to the Equator is important.

Q'enqo's Intiwatana illustrates this phenomenon in a fascinating way. In this case, the two *sukankas* show both night and day at once; and together they represent our planet.

On September 23rd and on March 21st beginning at sunrise, we'll be spectators of these multiple effects of lights and shadows. At that moment the Sun's light, cut by the small stone wall (which I already referenced when describing the June solstice), will follow a path up to the point where its rays allow half of the Intiwatana to be fully illuminated, leaving the other half in darkness. Day and night are manifested as equal parts in the astronomical evidence produced in Q'enqo's Intiwatana. After this symbolical division of the Intiwatana in light and shadow, representing our planet, the phenomenon will continue until the illumination of the right of this instrument in a line produced by the fissure that exists on the wall. Advanced studies with solar initiations and more evidence concerning this observatory in relation to the equinoxes will be presented in other book.

The equinoxes are taking place. One of them, in September, will show us the way of the earth so we'll start preparing the fields for their respective roles, thus allowing life's continuity. The other equinox, in March, will indicate the time for harvesting the earth's fruits is near. Our ancestors smiled in their hearts because their forefathers had not been mistaken.

Sunrise during the equinoxes in Q'enqo.

The Sun with its projection of light, divides Q'enqo's Intiwatana into two equal parts, the fissure effect illumination may be seen on the upper middle part of the photo (Equinoxes).

The world divided into equal parts day and night, note the little line at the upper right part of the photo, line produced by the fissure that is in the small wall (Equinoxes).

Beautiful *chakana*-fountain. At sunset during December solstice (this phenomenon is visible during the Andean spring and summer), it can be observed how the ledge and the relief of the *chakana* (square cross) project their shadows outlining the form of a Priestess. Their hands match with the proportions of the falling water, as if receiving it, which creates an artistic phenomenon of esoteric beauty. Paying respects to the astral priestess in the fountain-*chakana* in Ollantaytambo. The picture above was made in 1998 and the one below in 2007.

OLLANTAYTAMBO, SANCTUARY OF THE WIND

Ollantaytambo is one of the most magical sites in the Andes. Its natural setting and sacred geography are exceptional. This land where mountains and rivers converge, produced seeds which germinated as great people and communities, among them the Children of the Sun. These ones who knew well the great truths of the Cosmos and the Earth, perpetuated their knowledge of sciences and technologies and leaving us with many impressive monuments.

Only recently, since the beginning of the new Pachakuti, has the voice of the Hamaut'as been heard again, 1992 signals great changes at all levels. But what is even more important is that: in the slow and decided approach to a more precise understanding of our own history, we see that it has remained dormant the past 500 years. Logically, life shows us that everything evolves, including man's thoughts and concepts.

In 1992, many gatherings took place at all levels of our society. The cultural sector expressed itself through publications of a variety of materials, but the emphasis was mainly toward the coming of the Christians to America, rather the coming of the invader and destroyer of a culture.

No one will have the last word concerning our past. The intelligent man, for his own development, will go on carefully looking for concrete evidence. The discoveries that will be made in the years to come will be the testimony of a different history than written by Westerners.

Satellite pictures that were taken of this part of the Andes, revealed that some of its important archeological centers followed precise lines, especially one called Wiraqocha's Route.

Wiraqocha's Route was rediscovered by Maria Scholten D'bneth with expanded studies by the architect, Carlos Milla Villena, as revealed in his work, "Genesis of the Andean Culture." From satellite pictures it can be amazingly seen that this Route connects with various pre-Columbian cities for hundreds of kilometers between Bolivia and Northern Peru.

This Sacred Path includes the Sanctuary of Ollantaytambo. The pre-Columbian people generally built their cities, temples and *wakas* on electromagnetic convergence points to guarantee that they would benefit from the highest possible vibratory frequency.

In the Inkamisana astronomical observatory in Ollantaytambo, there are five carved measuring stones that come out of the wall. On important days of the year, such as the equinoxes, these stones project shadows that take the shape of a human profile, looking like a master who is playing with a wind instrument. This solar instrument is a very sophisticated one and in some future publications will be revealed more of its functions. See photo and artwork.

Pinkuylluna mountain in Ollantaytambo.

The measuring stones of the Inkamisana astronomical observatory project shadows that coincide with their horizontal and vertical edges. Furthermore, the center stones produce a profile shape, suggesting a master playing a wind instrument.

The musician of life. Artistic representation of one of the phenomena which take place on the equinoxes (Inkamisana, Ollantaytambo).

Ceremonial fountain in Ollantaytambo.

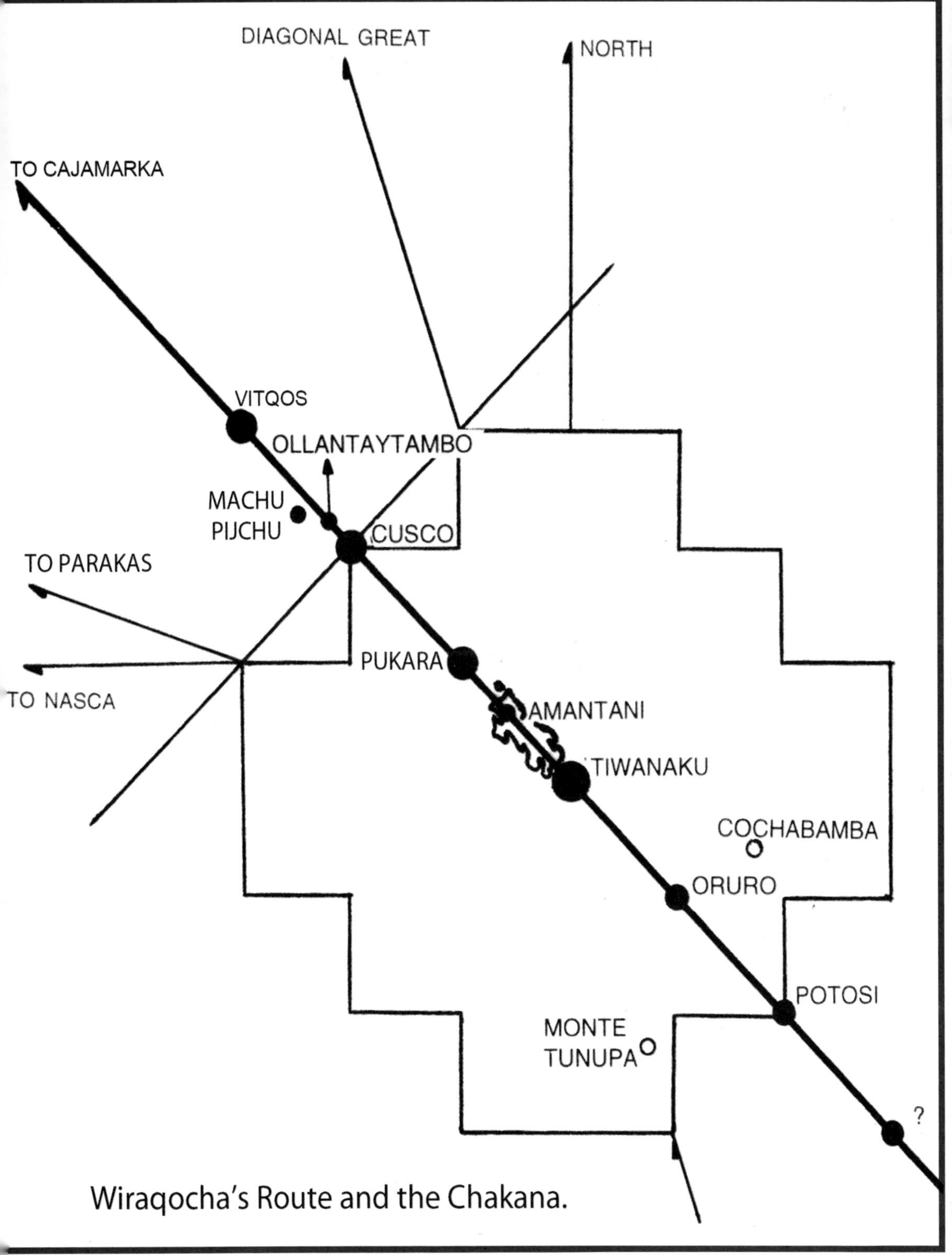

Wiraqocha's Route and the Chakana.

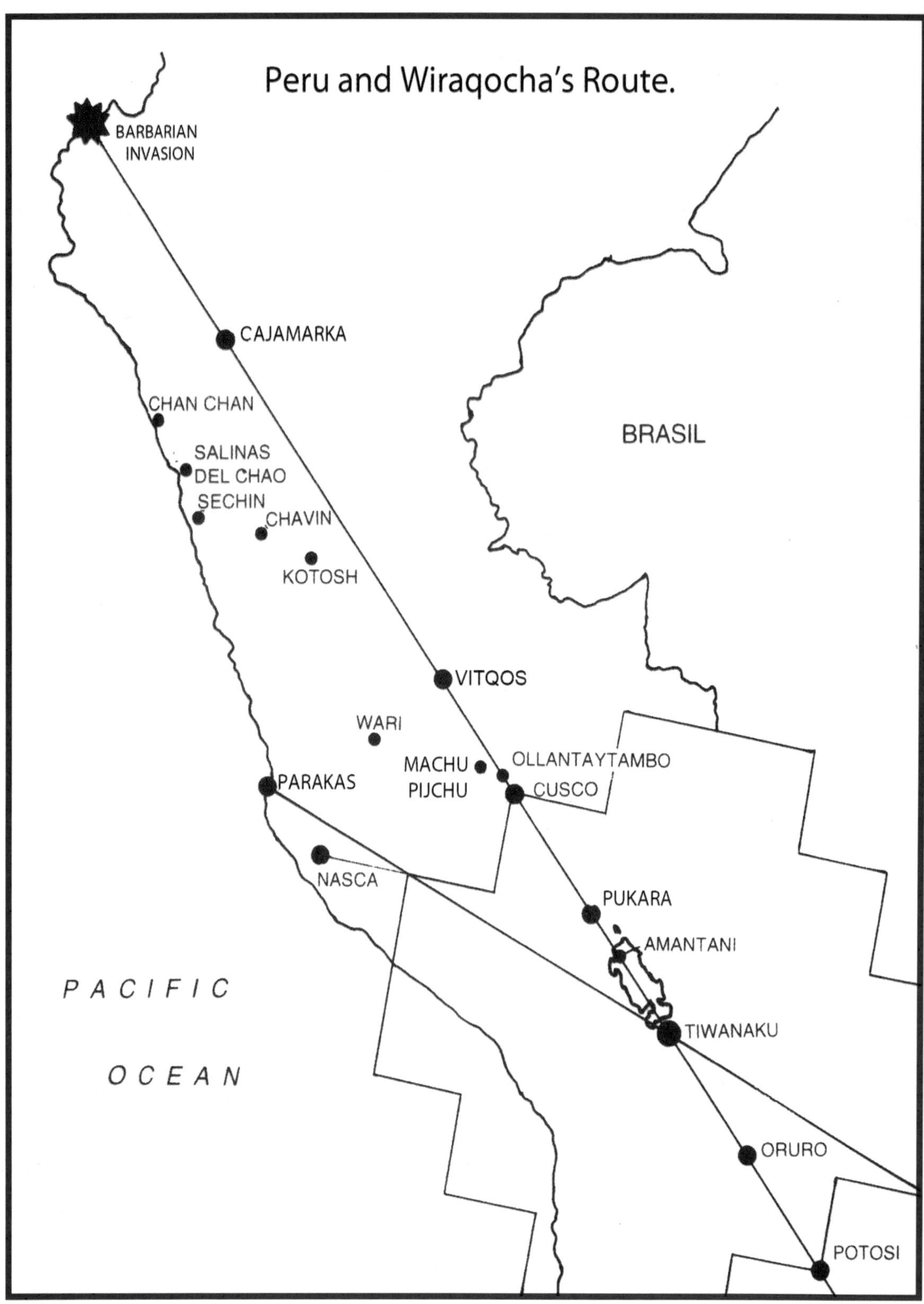
Peru and Wiraqocha's Route.
BARBARIAN INVASION
CAJAMARKA
CHAN CHAN
SALINAS DEL CHAO
SECHIN
CHAVIN
KOTOSH
BRASIL
VITQOS
WARI
MACHU PIJCHU
OLLANTAYTAMBO
CUSCO
PARAKAS
NASCA
PUKARA
AMANTANI
PACIFIC OCEAN
TIWANAKU
ORURO
POTOSI

PART IV

MACHU PIJCHU, CRYSTAL CITY AND CITY OF THE RAINBOW

1. MACHU PIJCHU, VOICE OF THE CENTURIES

Eighty-six years of history* is really too short a time to run out of ink, paper, and hope that someday we'll be able to precisely understand the meaning of the Sacred City of Machu Pijchu. There are hundreds of books that have been published on the subject and surely hundreds more which remain to be written. Throughout the twentieth century, we have recreated Machu Pijchu's history. There have been interpretations reflecting exacting studies as well as fantastic and incongruent stories. It has already been said that a thousand men, a thousand religions. A thousand men, also means a thousand ways of thinking.

So, within the framework of possibilities and with all due respect to those who have added their thoughts about the mysteries of that Sacred City, I am joining this beautiful and enriching journey to touch on a few points which up to now, have not been published or even thought about.

*I mention 86 years of history since its rediscovery in 1911, because this book was written and published in 1997.

Machu Pijchu, the Crystal City.

THE SACRED GEOGRAPHY

All the ancient cultures of the world have paid homage to the sacred aspect of nature. It was their main point of reference in the building of their temples, altars and palaces. Studies of both Eastern and Western cultures, show us that they generally chose the highest points of their surroundings. The mountains always were the guardians, where the protector entities lived, maintaining contact with the world of men and women through their spiritual representatives and oracles.

The Andean culture aligns in some aspects with the world's oldest societies, but it was different. In its technological development and construction, however, neither the highest peaks nor deepest valleys were obstacles to its expression.

Wiraqocha's Route, the Sacred Path projecting South-East North-West, is an example of the high technological level achieved by this greatly evolved Andean culture relating to and archaeo-astronomical perspective.

All along the Andes, there are fascinating examples of this sacred geography that was created along its route: Markawasi, the Pampas of Nasca with their designs; Kalakhumo's doors (North of Lake Titikaka); the Ajayu Marka altars and the Enchanted City (South West of Lake Titikaka); the Tukume desert with its pyramids; the lagoons of the Waringas in Piura; the Apu Waskaran in the White Cordillera; the volcanoes and canyons of Arequipa; Ollantaytambo in the Sacred Valley of the Inkas; Machu Pijchu with its mountains and valleys... and I could go on and on to list many more places where nature's genius is walking hand in hand with sacred geography.

For example, the geographical area around Machu Pijchu is among the most breathtaking. The Inkas' Sacred River, as it follows its course into the jungle's green canopy, describes shapes that remind us of the Amaru or Serpent. As it reaches the Machu Pijchu area, it adapts itself to those valleys and forms a very peculiar shape as the Omega of the Greeks: the shape of the Mother or feminine principle. It's this symbolic expression of Pachamama that shows us, in a spontaneous way, the best place for Initiates to reside.

The city of Machu Pijchu is situated at approximately 400 meters above the Willkanota River, nestled in a central point surrounded by mountains. In the East, directly beyond the Willkanota River, the observer will notice on the left side of Putukusi Mountain, the stair-step symbol repeating itself naturally and quietly on several levels. These stairs as a symbol on this mountain create the energy shape of the Andean Square Cross. This same symbol can also be seen in its natural state in other important mountains of the Andes.

Several researchers have already written books about the relationship between the region's most important sacred places and Machu Pijchu. Alignments link the snowy mountains with important sites such as temples, altars and places where astronomical manifestations occur. It's only been recently that we have been rediscovering this effect.

During the pre-Columbian era, the Andean peoples' cosmic-vision and ideological conception of life allowed them to fully integrate themselves with their surroundings. For them, the mountains were like natural pyramids, places of great magnetic concentration. It is this magnetism or power that the Andean people felt and recognized as the energetic force that resides in these high realms, which they called Apus or Achachillas. The Apus are the places where our ancestors assembled and where they still assemble today.

Machu Pijchu is often wrapped in a shroud of fog under a low ceiling of clouds, especially during the rainy season. Geographical and climatic conditions influenced the people of Machu Pijchu in the placement of the representation of the Apus in their temples and altars. This presence was marked by stone sculptures that looked like their revered mountains. Obviously, this was not just an artistic expression, but a different concept of spiritual recognition of the power entities who were part of their geographical surroundings.

Sunrise in Machu Pijchu
and Putukusi Apu.

FOLLOWING THE MASTERS' PATH
A PILGRIMAGE TO THE CRYSTAL CITY (...MEMORIES)

"The hardest part was to take the first step, but to Dare was one of the sages' great teachings. This inner impulse had to be brought to expression to take us to the other reality that would then be our birthright.

The paths had been traced following a superior plan. The path which leads to the Crystal City or to the City of Light, was destined for all the children of Pachamama who honored life, walking in the state of Tukuy Munayniyuq or beings of Pure Love or Unconditional Love.

The City of Machu Pijchu was the Crystal City. This great Crystal, along with its sacred geography, was the place of all changes and eventually all Initiates of the kingdom would have to walk on its sacred ground and honor its Apus, in memory of their ancestors.

The first step in this pilgrimage is our first step of ascension. Willkamayu, the Sacred River, would accompany us with its many voices, teaching us to be like him who never is the same, but always is one. Willkamayu's voice (Willkanota) would reach our veins and the mountains of the Sacred Valley would send the great Titi (Puma) so that it would walk with us and transmit its valor, security and trust in life, necessary for this long path that leads to the Crystal City of Machu Pijchu.

The path is marked, but would be inaccessible to the profane. Along the way its springs and waterfalls would be like the lights of remembrance of the words of the Great Amaru or Sacred River. The birds, perched in the trees, would repeat to us what the Apus said. Our backs, heated by Inti's rays (the Sun), would remind us of our inner being's great metamorphosis and the Sacred Kuka Leaf, which remained behind, in the shape of a K'intu, would remind us through its aroma, that it had once been Pachamama's Privileged Daughter. Pachamama offers us her juices and nectars to nourish us along the way, helping us understand that this was and still is human's spiritual food. Along with Mama Kuka, Mama Sara (Corn) would also accompany us, preparing the physical food that the master calls toasted corn to give us the necessary energy so that with will-power and harmony, we could bring our material offering to the Sacred City.

Ausanqati and Salqantay Apus would guide our way and in the dark of the night, the stars' dance foretelling our passage. Llamaqñawin (Llama's eyes) along with the *Chakana* (Southern Cross), with their vertical eyes and brilliant stars, will be the guides so the disciple of life could follow this ancient path toward initiation, with honor and decision as long as the Children of the Sun will last.

Finally after having crossed several peaks, valleys and narrow passages through which only the Puma can go, we would reach the Temple of the Water and of the Wind... and there Wiñay Wayna, or City of the Eternal Youth, would be the last resting place on our pilgrimage. On the following morning, long before dawn and the first bird's song, this student of life would be awake and saluting the spirit of the water, having already prostrated himself to the ground to listen to the words of silence. Touching his heart, they would tell him my Child, Service is the Being's reason for living... go on your way. With a moist face and a beating heart he would approach the Door of the Sun (Inti Punku), to salute the Apus with deep reverence and with tears on his cheeks, realizing that he was nothing more than a leaf in this great tree of life, always moving within Pachamama. Meanwhile Inti, the Sun would have rippled the horizon, touching the roofs of his snow-capped mountains in a humble gesture. Thanks to these Apus' doors, Inti's light and heat would murmur with each moment of life and only with the awakening of the Qori Q'ente would announce with its song, the emergence of a new Inti Khana or Being of Light.

The Crystal City, the City of Peace, the City of Light, Machu Pijchu will always be a cradle for the new man and woman." The following photographic sequence shows us how Andean people saw nature's spirits and its sacred geography. Machu Pijchu, in this case has been a very special place for the rich abundance of nature's spiritual manifestations. Without doubt these sacred surroundings were the appropriate site for the Andean masters to settle among these mountains and build their Sacred City.

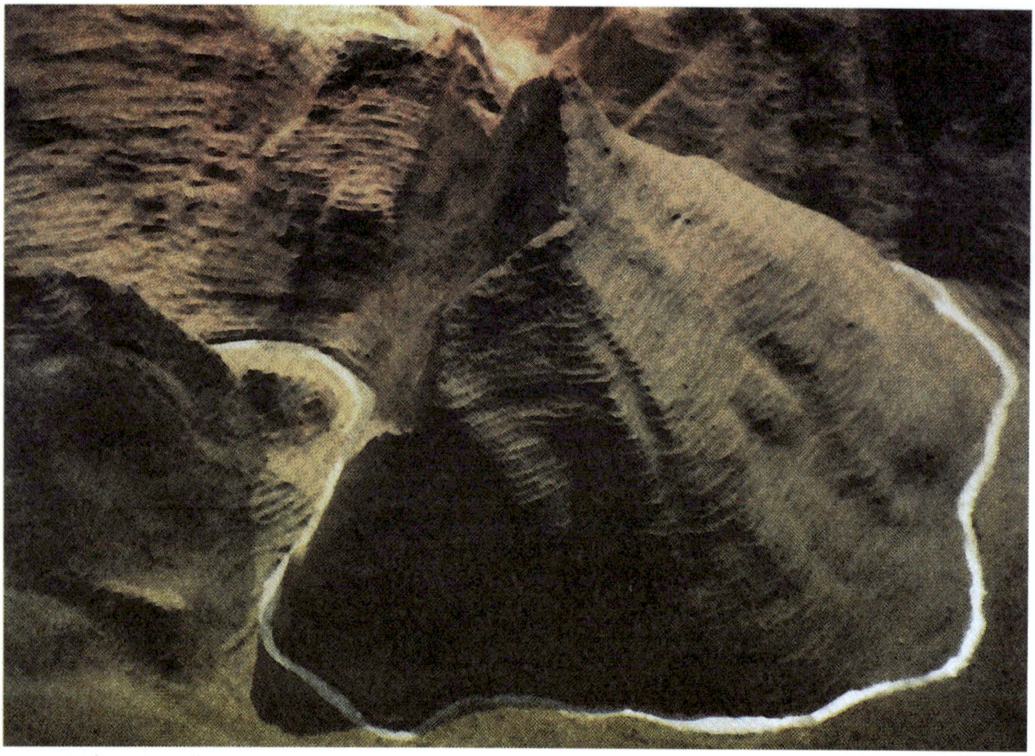

Definition of the Inkas' Sacred River around Wayna Pijchu and the Crystal City of Machu Pijchu. Symbol of Pachamama.

The symbol of Pachamama surrounds the Crystal City of Machu Pijchu.

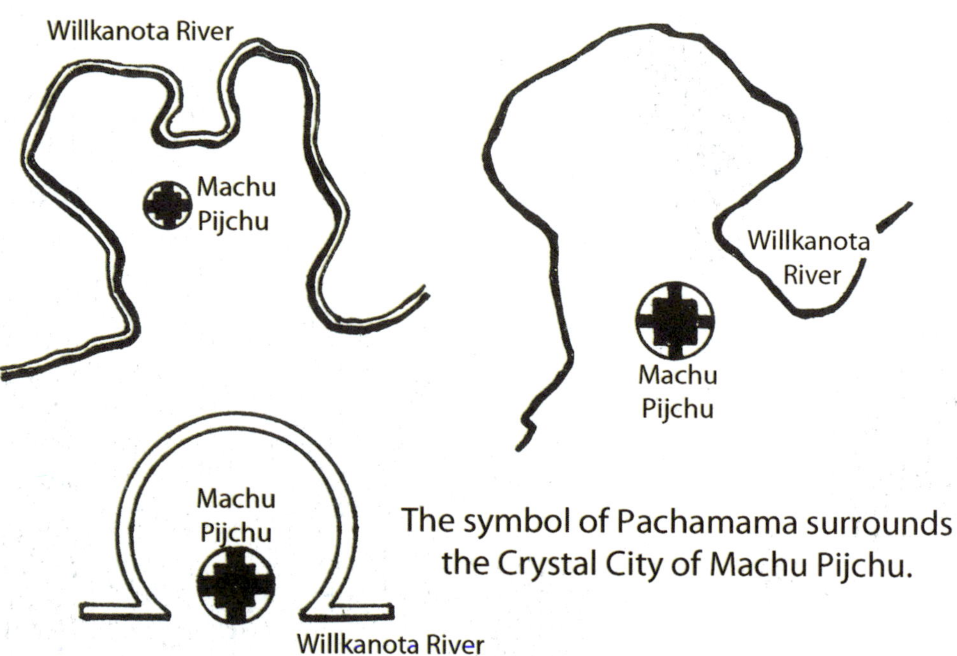

Machu Pijchu and its sacred geography.

2. WIRAQOCHA IS COMING BACK AND HIS WAKAS ARE AWAKENING

If I mentioned each and every place along the Andes, my use of ink and paper would be unending. After visiting some *wakas* and temples, we arrive at Machu Pijchu. This Crystal City or City of Light and the Rainbow, is like an open book in which each page presents a passage with an open plaza and the stairs leading to it. We want to read this book written with eternal stones, turn one more page, and start a new chapter, the chapter of the Masters of the new era.

Some scholars agree that astronomy was a very developed science to Andean people and the Inka culture left us with evidence of this science. In their desire for perpetuity, they built the magnificent city of Machu Pijchu. Once again, Pachakuti, the initiate, was its principal visionary.

Magic geography in Machu Pijchu.

THE CHAKANA AND LLAMAQÑAWIN

Known as the Southern Cross and The Llama's Eyes, these constellations walk together to announce the cosmic cycles and the advent of the water as a renewing and cleansing symbol. They follow their path between the equinoxes or solstices as predecessors of stellar phenomena.

The constellation of the Southern Cross or Chakana is now visible in the Southern Hemisphere. Centuries before the coming of Jesus Christ, it appeared a little bit more to the North and could be observed on the Southern horizon of the Mediterranean. This led the Greeks to associate it with the constellation of Centaurus and its stars Alfa and Beta which are aligned with the Chakana and are called "Llamaqñawin" (Llama's eyes in the Andes).

The constellation of the Cross, was the main indicator for the cardinal points with the South being the easiest direction to pinpoint. To do this, all that has to be done is to trace a line from the upper star, projecting it through the opposite star, pointing South.

At the end of the Fifteenth century, Vasco de Gama discovered the Southern Cross while navigating to the South of Africa and from 1603 on, the Cross was included in Western catalogues, thanks to the intervention of the astronomer J. Bayer.

The constellations of the Llama and the Cross preceded the celebration of the new solar year. The Christian Cristobal de Molina mentions that *"when the month of May arrived they came out in a procession with two llamas of gold and two llamas of silver and they would repeat these acts in memory of an ancient time..."* In a way, these celebrations were the introduction to the Inti Raymi, in June solstice. In the month of May, the Chakana reaches its highest point and seems to be vertical, with each one of its four stars signals the coordinates almost perfectly.

The church tried to erase the ancient celebration of the most holy cross in May, which had celebrated the presence of the Chakana reaching its highest point during the first week of May.... *"For the Northern Hemisphere, the main reference is the Polar Star. This star seems to be alone in its surroundings, with its solitary celestial presence originated the feelings and mentality typical of the Northern Hemisphere. And on the other hand, in the Southern Hemisphere, the presence of the Chakana's four stars, evolved in a more collective and harmonious way..."* (Carlos Milla Villena, Genesis of the Andean Culture -1983). In this way, the Chakana's horizon reflected in all directions, thus turning the Children of the Sun into individuals with a magical perspective of the world.

Fortunately, Machu Pijchu, the cradle of light, recorded the passage and position of these constellations in its stones, its structures projecting light and shadow to show us the way.

The Chakana's Table or Offerings' Rock.

THE CHAKANA'S TABLE OR OFFERINGS' ROCK

Also called Sacrifice rock, it is located a few meters from the original entrance check point, and is an observatory or *atalaya* in the high agricultural sector. The Inka trail practically takes us to that section, before entering the Sacred City.

Since 1911 many investigators have passed through Machu Pijchu, simple people often unaware and insensitive; but who did identify themselves with our culture... the minority. It has been suggested that many stone artifacts were taken from the sanctuary; others were destroyed. Others recently created in the reconstruction of the site were placed without any cultural or historical reason. My hope is that time will be our judge, and that the displacements of material evidence such as stones and walls were deeds intuitively guided by a person's subconscious impulse. It is often our inner unknown capacities that show us the right path and give us the necessary intuition to make the right choice.

But whatever the case may be, these stone works and their geographical placement show us the level of mastery our ancestors attained in their time.

In 1974, Arminda Gibaja Oviedo excavated an area in the Roca Labrada (Carved Stone) sector with her team never founding any evidence that it could have been a cemetery. On the contrary, it was demonstrated, as Alfredo Valencia Zegarra says, that *"…the presence of eroded boulders from the river and other regular size stones able to be transported by one person, allows for the idea that these had been brought there from the river's bank, as a sort of rite similar to that of the apachetas, which is still practiced in our traditional popular culture today."* He also adds that *"…it has been established that the rock, which has a hoop on its North East extremity, was transported to the place where it is actually located showing it had a specific ritual purpose."* (Valencia and Gibaja 1992:213), (Notes taken during the International Archeology Seminary-Workshop on the National Historical Sanctuary of Machu Pijchu as World's Patrimony; Cusco 1993).

This coincides with the idea that the Roca Labrada is there for a ritual purpose and that its East-West orientation is one more indicator of its importance at a stellar level. Furthermore, observations made on key calendar dates give us astounding results. On May 2nd, there is the celebration of the cruz velacuy or May's Cross, a feast which was instituted by the church with the purpose of eliminating a very ancient cult that honored the Southern Cross when it was in its highest position in the Southern sky. The sign which appears in the Roca Labrada or Offerings Rock is the introduction to the arrival of the Solar New Year, which starts on June 21st (solstice), the original day of the Inti Raymi, which the church moved to June 24th to make it coincide with the Christian feast of San Juan, a widely celebrated holy day in Western countries.

At dawn, on May 2nd (celebration day of the Chakana's feast) as the Offerings Rock is touched by the Sun on its Eastern side (the part which has a triangular, pyramidal shape, that looks like a mountain peak), a shadow is projected to the West. As it moves backwards, it fits perfectly in the opposite corner of the stone, thus creating the appearance of a hidden Cross, formed by the shadows extending in four directions. In the ceremonial aspect, the Chakana's Table is the Rhombus (sacred Andean structure) which contains its offering. The shadow projected by the pyramid structure of the Offerings Rock toward its opposite corner marks the movement of the Sun, like a dark needle. It is announcing at the same time the prompt coming of June solstice.

On May 2nd, this upper triangular shape forms a shadow on its table after the rising of the Sun. This shadow looks like a Cross and its extreme point indicates the Western corner of the table.

I reproduce in the following text the order of the months according to the chronicler Guaman Poma de Ayala:

January, Camay Raymi
February, Paucar Uaray
March, Pacha Pucuy
April, Inca Raymi Quilla
May, Aymoray Quilla
June, Cuzqui Quilla, Inti Raymi
July, Chacra Conocuy
August, Chacrayapuy Quilla
September, Coya Raymi
October, Uma Raymi Quilla
November, Aya Marcay Quilla
December, Capac Inti Raymi
(1993. Fondo de Cultura Económica - Perú)

Now, we will see how these months relate to important astronomical dates:

1. June: Inti Raymi, New Solar Year (June solstice). Month of abundance and celebrations.
2. July: Month of abundance, and the end of the earth's resting.
3. August: Month of the Wind (Wayra).
4. September: Qoya Raymi, time of purifications (equinox).
5. October: Uma Raymi Killa, celebration of the beginning of the rainy season. At the end of this month the constellations Llamaqñawin and Chakana begin to disappear.
6. November: Aya Marq'ay Killa, meditation for the dead. More planting continues.
7. December: Qhapaq Inti Raymi. New Earth Year (December solstice). The rain is nurturing the earth.
8. January: Month of resting, time of scarce nourishment. Season of rain.
9. February: Llamaqñawin and the Chakana (Alpha and Beta stars of Centaurus, and the Southern Cross), return to the Earth. It continues to rain.
10. March: Beginning of the harvest, end of the rain (March equinox).
11. April: The harvest continues.
12. May: Aymoray Killa, end of the harvest, celebrations to the Chakana (Southern Cross). Preparations for the New Solar Year to come in the solstice.

THE CROSS AND THE LLAMA

In the Sacred Plaza, Sacred Square, that we can call The Haukaypata, there is a distribution of spaces, altars and stones with important references. The shape looks like a square cross, with the inclusion of the three open windows located on the East side. The semi-temple with its main altar, is to the North, the priests' quarters to the South, and the great open space with its semi-circular tower to the West.

In the area within the enclosure of the three open windows, used astronomical purposes, there is a flat stone with the well-known symbol representing three levels, showing us the ideological conception the Andean people had in relation to their universe. This three-leveled symbol is the result of the evolution of the Southern Cross in its celestial movement. That constellation was used as a basis for the creation of the perfect multi-leveled symbol that became known as the Chakana or Andean Cross.

In the direction of the West and on one side of the access path to the Usnu of the Intiwatana, there is another stone that gives the impression it is coming out of Pachamama. This stone has been identified as the Southern Cross, but if it is observed quickly, it loses its stellar meaning. Its extremities, which are supposed to represent the four stars Alfa, Beta, Delta and Gama Cross of this constellation, are completely reversed. Observing the position of that carved stone representing the Southern Cross, we would be surprised to see that the North would not correspond to the North and the same thing for the South. In this very special case, this position is due to an ideological concept, since the Andean people saw, how in their cosmology, the Chakana was always accompanied by Llamaqñawin or the Stellar Llama. During the month of May, the constellation of the Chakana is in its highest position and Llamaqñawin is accompanying it. At the end of October - November, before the rainy season, both constellations almost disappear from the evening's horizon and from January - February they slowly come back, to repeat the continuing cycle.

Two months, May and October, are still well-remembered and celebrated today. May is the month of the crosses, with ceremonies all over the country. On the other hand, the celebrations in October have been reserved for very specific places. One of the communities that most remembers October as an important announcer of the rain as the sky's messages is San Pedro de Casta, near the power center of Markawasi.

There is a myth narrating the movement of the Chakana and Llamaqñawin in a very special instructional way.

"... The Great Stellar Llama (Yaqana), whose body is black, comes down to Earth every night to drink water from the springs and ocean. This is to prevent the water from rising too much, causing a flood. In this way, the eyes of the Yaqana have saluted the Earth just before the rainy season starts. As it continues drinking the ocean's water it allows the rains to come down again. When the Llama raises its eyes and starts to urinate on the earth, it starts the rainy cycle again for people's greatest benefit."

In 1993 I discovered that in the early hours of the morning after the rising of the Sun and for a few months of the dry season, this stone representing the Southern Cross, will project a shadow reproducing almost perfectly the silhouette of a Llama, which will grow shorter as the Sun progresses in its course. This phenomenon is even more surprising because of its presence and the symbolic and magical representation that this Llama offers to the Cross. We'd say that Llamaqñawin always follows the Chakana. A clear night away from the city lights allows us to see in the South by the left side of the Chakana -a dark area the Andean people call Llamaqñawin, where the Alfa and Beta stars of Centaurus constellation represent the Eyes of the Llama. This manifestation parallels the dry and rainy seasons. During the rainy season, it will be impossible to clearly see this phenomenon. First because of the position of the Sun, and second because of the rains that prevent a systematic observation. Let's not forget that the Stellar Llama in order to "drink" the Earth's water must disappear for a period of time.

The study of the proportions of the Southern Cross allowed the creation of measures and dimensions, which were widely used by the Andean people from before the coming of Jesus Christ. Along with measurements, multi-leveled structures were built that maintained continuity with the stars.

The multi-leveled symbol was used to define the different *pachas* (dimensions or energy levels). The Andean projection did not only limit itself to the three levels of the classical Chakana, it went further and created greater Chakanas with seven levels. With these expressions, we can clearly see that the Andean culture was dynamic and creative and its limit was not in the projection of three, four or five, but reached the infinite of our Universe. If we go for a walk in the area called Chakan, we'll see symbols with more than three levels, and if we go down by the same little river, we'll see multi-leveled symbols one of with seven levels. In Tiwanaku, in the Southern part of the Sacred Lake Titikaka, there are also Chakanas with more than three levels.

Chakanas with two, three and four levels (Tiwanaku).

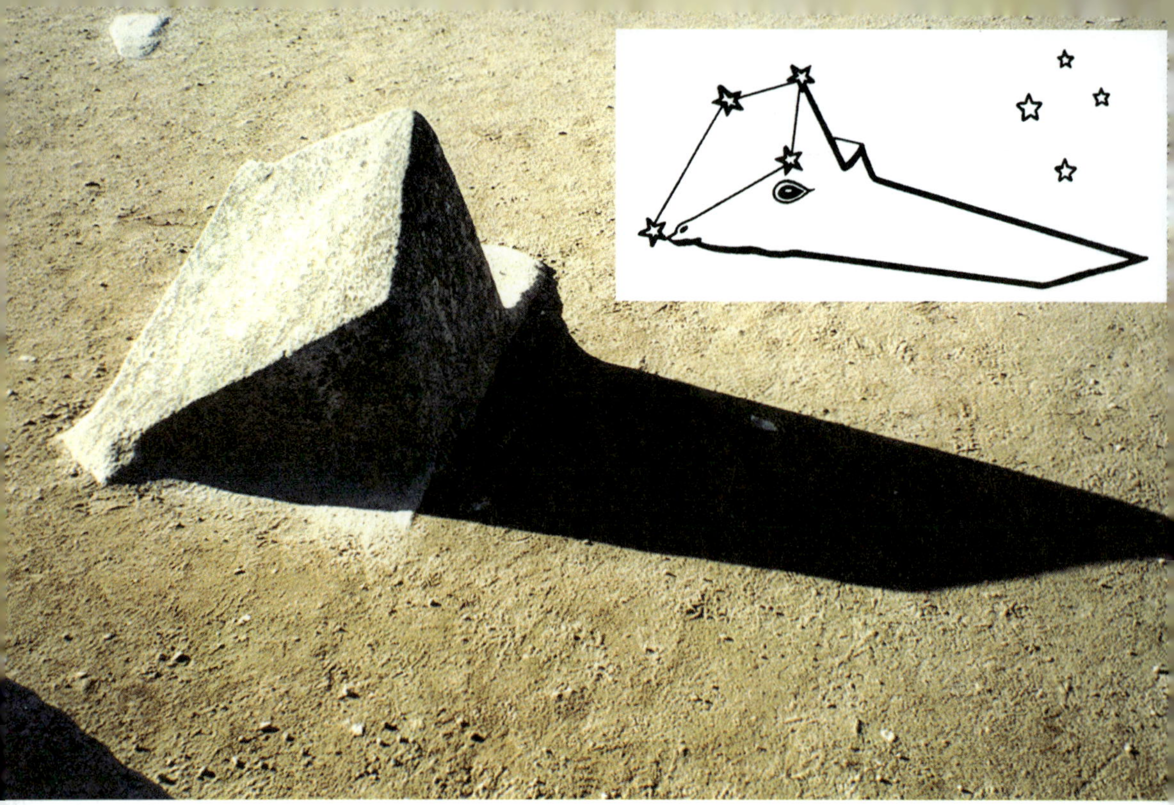

Chakana and Llarnaqñawin. The upward face of the stone representing the Southern Cross is illuminated and at the same time its structure is projecting a shadow which outlines almost perfectly the darkened silhouette of a Llama's head. The Chakana and Llamaqñawin seen in a stellar mirror. Artistic reproduction of the phenomenon in which the Llama always accompanies the Southern Cross (Sacred Plaza, Machu Pijchu).

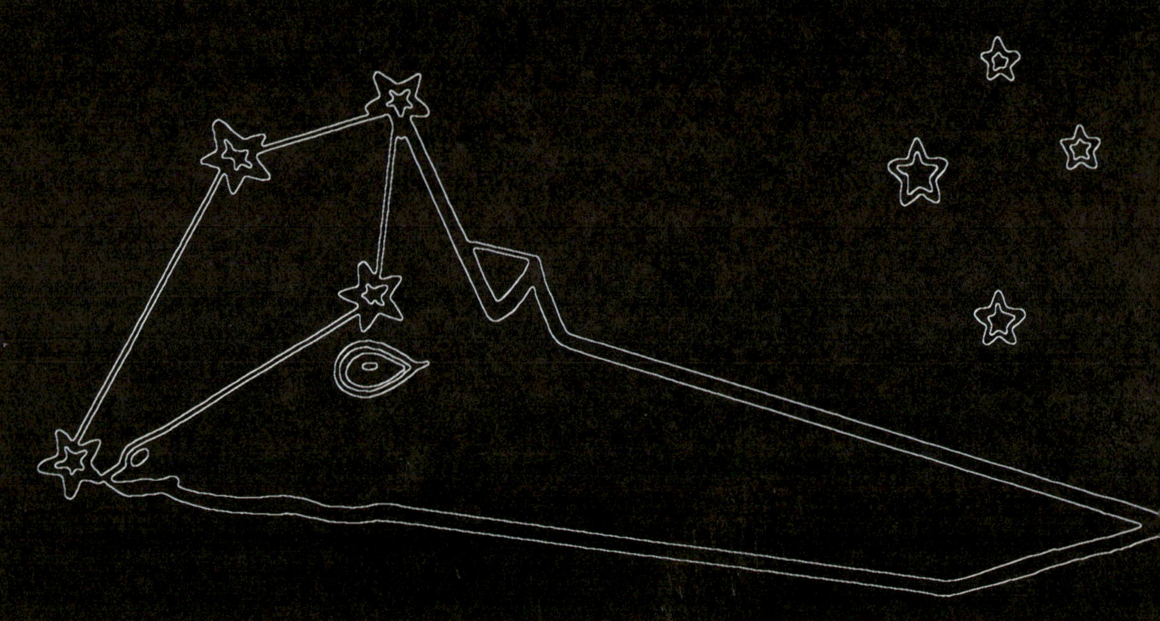

Sunrise and the Sacred Square with the Three Open Windows Temple.

The Intiwatana, solar observatory and Wayna Pijchu Apu.

3. INTIWATANA
THE INITIATES' OBSERVATORY

Where the Sun touches us within our hearts and where the Sacred Puma awakens.

After visiting some power places, we reach the part, that because of its placement and stone-works, has inspired all who have had the opportunity to see it. I am referring here to the Intiwatana and all that surrounds it.

This stone structure is probably the most important of the sanctuary, because of its dimension and shape. Numerous interpretations have been made concerning the Intiwatanas and its use. Apart from the local folklore interpretations, I can assert "that our ancestors did not give priority to daily time control, such as the prolongation of fractions like the seconds of a clock, that are so necessary in our current complex life."

I have already made reference to some of the functions of the Intiwatanas. In this particular case, we have already shown that the table serving as a base for the prismatic structure has four angles oriented according to the cardinal points (South, West, North and East). In one special case, the stone measuring rod extends toward Wayna Pijchu Mountain, pointing directly to the North.

The whole of the Intiwatana, as observed from its surroundings, presents itself as the crown of a truncated pyramid, which has been completed by a system of terraces. So, the greater structure of the Intiwatana is part of a natural hill that was worked in situ.

Many times, people have tried to give interpretations of the functions of this Intiwatana. Up to now, no one has been able to provide conclusive evidence as to its meaning. Conceptually, most agree that the answer lies in the interplay of light and shadow. It is through this projection of shadows and light that the coming of the equinoxes and solstices can be observed. I wish to add my observations and discoveries made during several years concerning the best time to observe these effects of light and shadow which are during the solstices and equinoxes.

The Intiwatana and rainbow.

THE PRIEST'S OR PRIESTESS' TABLE

This is an apparently insignificant table, situated to the West of the Intiwatana by which one reaches the South-western terrace. It is slightly inclined by a few degrees and adapted to the great rock located on its right. On that apparently simple table I was able to discover in 1992, valuable astronomical evidence.

In order to see what happens there, let's go back to the most important day of the Andean calendar; June solstice, the day of the Inti Raymi.

The Ceremonial Table and priestess saluting to the ancestors.

The Solar Initiation.

Initiation Staircase with Seven Steps that guide us to the Puma's ceremonial table and to the Intiwatana's Usnu.

"The priest or the priestess, after his/her respective invocations and fasting, has walked the Amaru's path and has arrived half-way up the staircase which leads to the Ceremonial Table. The sound of the *pututu* (sacred instrument) and the birds' song have already announced that very soon, another one of Pachamama's Children will become a Son or Daughter of the Sun."

In the exact moment, the Sun illuminates the upper part of the Great Pyramid, and the Intiwatana's Ceremonial Rock's power is once again revived by the contact of the Sun's force with its center. **In this eternal instant of high magic, the Intiwatana's great stone prism lets part of Inti's force pass over its top to illuminate The Forehead of the Priest or Priestess who remains motionless during this transcendent moment**. The Sun has touched the priest on top of the head and Inti's force has gradually gone down to his Forehead to initiate him in the Being's higher levels, to make the Munay or Unconditional Love germinate in his heart. The Golden Q'ente (Hummingbird) is now present and a colourful halo has already surrounded the priest, thus transforming him into a Lupihaqe or Intiruna (Solar Being).

In this eternal dream, the priest/priestess, surrendering to the Sun's arms, will now take part in the Initiation and Awakening of the Feline totem that is ready to happen.

The Sun's Peace has illuminated the priest's face and its heat has submerged him into full communion with Pachamama. Slowly, he has started to feel the Puma's breath and as Inti's presence reached his throat *chakra* (energy center), he has heard his inner voice saying Wiraqocha Wiraqocha, Wiraqocha. The priest, his hands resting on the Ceremonial Table, has started to caress it, having felt that his great friend and brother, the Solar Puma, was already starting to awaken (the two small concentric circles which appear to the South of his right hand are the Puma's eyes). Now comes the moment for the Puma to walk side by side with his Human Brother/Sister. Together they will accomplish the Solar Purpose.

Finally Inti has reached the priest's/priestess' Love Rose (the heart) and in a measureless explosion, life has illuminated itself.

The Puma has awakened and the Sun, with the help of his stones, has defined a triangle, thus illuminating the feline's eyes.

In 1992 I discovered how these two small circles located on the South side of the priest's/priestess' right hand, are illuminated each June 21st. The Sun, thanks to the presence of the Intiwatana's prism, and the rock to the right of the table, makes a luminous triangular shape over the two circles. This triangle is a sacred constant in sacred geometry, which in this place, shows us that the Sun needs to complete a revolution of approximately 365 days, in order to once again return to the same point, the Two Concentric Circles!

This astronomical phenomenon is one of the most astounding evidences due to its precision and magical effect. I call it a first class astronomical phenomenon, since every June 21st, the two small concentric circles are illuminated on that Ceremonial Table, thus giving a sample of the Sun's passage and movement.

Along with all of these astronomical phenomena occurring in different places in the Andes on this day, here in Machu Pijchu there also exist altars and temples built for this purpose such as the windows of the Solar Tower in the royal area. At the moment of the rising of the Sun, this Tower's Window, oriented towards the June solstice, allows Inti to come in and touch its Power Being or Sacred Animal. In a synchronized way, the different Hamaut'as or Initiates of the kingdom of the Children of the Sun would verify that their forefathers were not mistaken.

We are now in the fifth year of the New Pachakuti (this book was written and published in 1997). We shouldn't be surprised to see that we still have much to discover about our ancestors. During the 90's we were just a few celebrating these astronomical days, nowadays this is a public event and many people join this great moment.

Let's see the photographic evidences of these phenomena in the following pages.

Sunrise in June solstice, view from the Intiwatana in Machu Pijchu.

The priestess receives the Sun's light on her forehead as receiving the Lightning. The initiation, directly given by Pachamama and Wiraqocha, has begun.

Priestess in the process of Solar Initiation (June solstice).

Astounding initiating astronomical phenomena. The illumination of the eye of the Solar Puma occurs during the June solstice when the Sun touches the HEART of the priestess and a frame of light produces a triangular geometry with two small concentric circles visible inside. I discovered this in 1992 and see it as an extraordinary astronomical projection since every June 21st it returns with precision and ideological effect.

Ceremonial table and its illuminated eye, in the context of Machu Pijchu. Solar Initiation ceremony in the day of June solstice. Before 1997 pilgrims had the unique opportunity to be in this magic center peacefully.

Illumination of the Puma's eye. Fifth Level, initiation path.

HERESIES OF THE HOLY INGENUOUSNESS

From the point of view of the official critics, the existence, organization and evolution of the Andean cultures have always been marked with the presence of blood, holocausts and sacrifices (even human sacrifices?).

We understand that humanity represents the highest Level of evolution on our planet. Within very limited margins, man can do or undo things according to what seems to be right for him. I don't say according to what is convenient for him, simply because man regularly does things that really are not convenient to him.

If we want to take a few existing examples, sooner or later we'll get to the Andean culture, of which the last evolutionary period was characterized by the presence of the Inkas. Critics throughout the world agree that the Inka society was the most evolved we had known up to then. The Inkas, although they fought wars and made communities disappear from their original land, as they progressed with their expansion process, often used persuasion as their favorite means of conquering. Their message of persuasion was so convincing that very few villages or nations resisted. On the contrary, they accepted the new organization and the benefits of the new lords. The Inkas formed one of the greatest kingdoms of the pre-Columbian world and one of the best-organized society histories has ever known.

The organization of this society attained such a level of harmony that the people's basic needs, of nutrition and housing, had long ago been overcome. The people's hunger was controlled to the point that to die of hunger was impossible, as well as to die abandoned for lack of housing.

Often, the Inka kingdom is criticized as a society maintained with slave labor. But is it so terrible if those "slaves" considered their slavery as a divine condition, because the Sun provided them with their daily bread and their shelter for a decent life? If eventually the Inka people -inhabitants- seemed like slaves, it was because they submitted themselves to the earth, which as a mother, was providing its children with everything they needed.

HUMAN ... SACRIFICES?

In the studies I have made, I was able to prove that the Andean culture and along with it, the Inkas, had a very wide and precise knowledge concerning the stellar movements, climatic manifestations and other natural phenomena. "They knew the precise moment to start preparing the earth for its cultivation, since Lord Wayra (the Wind) had blown to the four winds, the path of Llamaqñawin (cosmic Llama). They knew that the Sun would not let itself be seen for some time, since Llamaqñawin was urinating on the earth, thus maintaining the rainy season and feeding the ground. They knew that the Earth was facing the Sun and that it would produce twelve hours of light and twelve hours of darkness, and that soon the rain would stop to start the harvest. They had seen that Llamaqñawin was upright, when observing the Chakana. Soon the Solar Puma would awaken so as to maintain the cold and dry period, and the rains would only come back in the appropriate season."

Those of us who live in the Andes, know that the rainy season and the dry season are opposites and that they have their precise times of existence. We say that the rains walk with December, January and February, but we also observe their presence in November and March. The absence of rain is definitely observed in June, July and August, but May and September also accompany this dry season. Even the simplest villager or most careless peasant would understand the climatic differences between the wet and the dry seasons. In our present time, technology and the rhythm of life we lead has made us a little bit "nature blind" and dependent on a system we don't understand, but still, we know that between the dry season and the rainy season, at least six months will go by.

Our ancestors knew perfectly well about these rhythms and they also knew there was no reason to fight against the current. By this, I mean that within our planet's ecological and biological systems, everything has its reason, and its own time of existence. This is the process where the dry season allows the earth to rest and prepare itself for future seasons.

Imagine that we are the astronomers of the village where, year after year, we have managed to verify the passage of the Sun by the Intiwatanas and have proven that the seasons (solstices and equinoxes) are signaled by certain effects of light and shadow with a great variety of magical meanings. In that way, it was possible to scientifically measure time. Nothing in this world will have the power to modify this stellar motion of the Earth around the Sun and as a consequence All of its other Repercussions.

If we make an abstraction of these phenomena and try, in an ignorant way, to "Go against time and swim against the current, trying to modify what Pachamama has established as Law: The Ecological Rhythm of Life..." thinking that our people are totally ignorant, then we may fall prey to a mentality that believes human sacrifice can change these cosmic laws. Then, perhaps, we would order it, as some of our history's well known chroniclers have mentioned: *"they made sacrifices to the Sun and gave in to the sacrifice called capacocha, during which they buried innocent children... five hundred innocent boys and girls buried alive standing-up..."* (Guaman Poma de Ayala). The purpose of the sacrifices was to guarantee the people's peace, to secure good weather, to foretell good omens, etc., etc.?

But then, tell me what is the reason the Inkas built so many solar observatories with the intention to scientifically study the movements of the Sun, to know our climate's variations and other terrestrial phenomena? If in an ingenuous way, some people will go on making us believe that the weather could change if one, a hundred, or a thousand virgins, or even a whole village would be assassinated... How could the killing of one hundred virgins make December (Rainy Season), turn into June and vice versa? Really one must be very naïve to believe that a culture as advanced as the Inkas could have maintained the practice of human sacrifices.

We understand that in the developmental process of a society, not everything is light. Executions and killings occur in a social context, and perhaps aren't always representative of that society's material or spiritual level of evolution. If that was so, we would have to say that throughout two thousand years, the Christians have sacrificed millions of people with no other reason than to increase their political power and material riches. If all of this served as an example, then our century would be a model of the most barbaric times we have known up to now.

When speaking of sacrifices, we have to be very cautious and know about all the variants of a culture. If an offering is being made, it is good to know why and what it is being made for. It's also important to know the right moment to do it or not and if it is needed in a material way.

The chroniclers who wrote part of our history, showed us in a descriptive way some of the things they saw. They were mistaken in their interpretation of some spiritual practices of the Andean culture, interpretation of an opposite dimension. This type of convenient interpretation, and documents written out of context, are typical of conquering cultures.

The architect Carlos Milla Villena makes a perceptive observation about human sacrifices. He says that the eradicators of idolatry are the origin of this devaluation of our Andean culture. Bernabe Cobo asserts that the majority of these *wakas* were temples for idols and many children were continuously sacrificed to those idols. If our Hamaut'as (elders) had written their colonial history in accordance with what they saw and heard, they would have narrated it like this: *"The invaders had many dark temples and decorated their altars with the gold they stole from us. The walls are full of tortured images and masochistic scenes. Their god was a dead man who was judged and whom they resuscitated through strange witchcraft so as to offer him again in sacrifice to his own father. During these obscure ceremonies, they all share the victim's flesh and blood."* It is obvious that a chronicle thus written would have been faithful reflection of what we would have heard the very same invaders say (Taipinquiri, The enigmatic Andean ethnoastronomy. La Paz, 1995).

I know that not everything was good about the Andean culture, but I am sure that there was more light than darkness.

So, to continue I'll transcribe what Victor Heredia tells us in a song called Taki Onqoy.

"THE DOOR TO THE COSMOS"

The door to the cosmos opened slowly
And there, Wiraqocha founded the world we see
All things and wild beasts and the civilizing cult.

The valleys and fruits, the beautiful meadows
And the water in a gesture of love.

Wiraqocha reigned amongst us, loved amongst us
And one early day went away
The God of Life, the God of the Earth
Crossing the waters of the sea.

Just as Quetzalcoatl in Mexico one day,
Both promised to come back this way.

My heart with its drum
Is beating at the doors of Tiwanaku.

My heart in its pain
Is calling to the armies of Tiwanaku.

The men who are coming are not Wiraqocha
There is no goodness in their actions.

Their magic is death, their love is for the riches
Of the Children of the Sun.

(Victor Heredia: "Taki Onqoy").

THE PUMA'S EPILOGUE

Our culture, the Andean culture, still is very little known. Our modern life has involved us so much that we can't even see the Sun anymore.

I am sure that from the Earth's core, Pachamama will make her voice resound and humanity will be the Being of Freedom, Love and Light. We know that all paths lead sooner or later to the Light and that if we don't go to the Light, it will come to us.

Only a few days ago, it was concluded that the pre-Columbian societies knew the principle of electricity. Ladies and gentlemen, we are facing the greatest revelations of our time and these will make, without doubt, an unprecedented revolution in our history.

We will still have much to say about the Andean culture, and the day that its children from the world become conscious of the fact that they are neither on one side, nor on the other, on the contrary, they are children of this land, of this Pachamama, their life will be different and they will become Children of the Sun.

We must understand our culture in order to be able to appreciate it at its highest level; it has not been perfect, but it has been the most complete of all. We have had a lethargic dream that has ended at last and the New Pachakuti is marking the beginning of a new cycle, a new Era of Light.

GLOSSARY OF QUECHUA TERMS UTILIZED IN THIS WORK

Achachilla: Our Grandparents, the "Apus." In some cases also translated as "apacheta" (site of offerings).
Ajlla wasi: House or dwelling of the Chosen Women. Where the Virgins of the Sun dwelled.
Alqamari: Zoo. (Polyborus chima chima. Phalcoboemus albogulares. Sacre) Bird of the family Falconidae, with white head and neck, body brilliant black and red legs. Also in brown color called "ch'unpi alqamari." Andean Cara-Cara.
Allpaqa: Zoo. (Lama pacos Linneo). Alpaca, paqo, paku, paqocha. American camel family. Bears very fine wool fibre of up to 70 cm in length. A very esteemed and respected animal among the Andean people.
Amaru: Serpent, snake, viper. By extension, rivers due to possessing a serpentine form.
Anka: Zoo. (Gerancaetus melanoleucus Veicillot). Hawk, eagle. Bird with head and neck covered with feathers.
Anti: Forested and Jungle region of the Andes. Region where the Sun is born. This word possibly originated the name of the Andes Mountains, the greatest mountain chain in the world.
Antisuyu: One of the four regions, or "Suyus" of the Tawantinsuyu Kingdom of the Inkas, corresponding to the East or "Anti." The Inka King Pachakuti dedicated himself to military trips into this area, possible partially conquering it.
Apacheta: Apachita, site of piles of rocks representing offerings that are placed on mountain passes or high altitude mountain sites. By extension, also referred to as "Achachilla."
Apu: Protector Spirit, or Guardian Spirit of a community. Dwells in the mountains and also in some important "waka."
Atoq: Zoo (Pseudalopex). Fox, canine, mammal of elongated face, with five toes in front paws and four toes in back paws. Beige, orange color. Habitat in highland plateaus ("punas"). Known for its astuteness and guile, and also for always following in the wake of the Kondor.
Ausangati: Awsanqati (Aswan aswan qatiq: the one who follows; aswan qatiq: he that can or is able). Beautiful snowcapped peak and main Apu of the Cusco region, reaches an elevation of 6372 meters above sea level. Located Southeast of Cusco.

Auki: Protective Spirit, mythical character inhabiting the high summits. Protective Being incarnated in the hills and mountains: the soul or spirit of high mountains.
Ayawaskha: Bot. (Banistería metallicolor Jun). Translates literally as "rope or vine of death." Plant with Magical powers.
Ayni: Return, recompense, loan, sharing, correspondence, retribution, exchange of actions or activities. In the Andes this practice still exists and is based on reciprocity.
Chachakumu: Bot. (Escalonia resinosa R. et P.). Chachakuma, tree belonging to the saxifragaceae family, has a gnarled trunk. Used in cabinetmaking, also as fuel in the highlands.
Chakana: Square Cross or Andean Cross. Stellar symbol inspired by the Constellation of the Southern Cross.
Chinkana: Tunnel, Labyrinth.
Choqe: Metal (from Aimara). Refined Gold. Any precious metal.
Enqa: Talisman, rock or stone in human or animal shape considered a totem.
Eqeqo: Represents a legendary Andean Master -Itinerant Teacher who travelled imparting knowledge and transmitting the message of Wiraqocha. Also known as Tunupa Wiraqochan. The starting point of his travels was the Altiplano Region around Lake Titikaka.
Hanan Pacha: The realm of the skies and the gods. The upper world of blue and white heights were the spiritual beings dwell. This realm is represented by the Apus, or Mountain Spirits, and its ruler is the Kondor.
Hucha: Crime, transgressions of the Law.
Illapa: Tellurian force represented by the Lightning Ray, the Thunderbolt. Light or flash produced by an electrical discharge in the atmosphere.
Inka: Monarch, king, emperor, sovereign, supreme chief of the Kingdom of Tawantinsuyu, leader, spiritual guide, Child of the Sun.
Inti: the Sun.
Inti Raymi: The festival of the Sun. In ancient times this spiritual gathering was celebrated on June 21st. After the coming of the Christians, this event of astronomical transcendence was arbitrarily transferred to June 24th and made to coincide with the Catholic observance honoring John the Baptist.
Intiwatana: "Hitching" post of the Sun. Raised part of a sundial that casts the shadow used as the Sun marks the completion of a year. Symbolically the point that marks the solar year. Solar observatory.

Kausay: Life, vital energy.
Kay Pacha: The present world, the world of here and now, of living beings, of the tangible and of all and everything that moves upon the Earth. It is represented by the Puma.
Khuya: A stone of power used for healing and protection.
Kuntisuyu: Qontisuyu. The western "quarter" of the Inka Kingdom.
Kuntur: Zoo. (Vulturs gryphus Linneo. Sarcorhampus gryphus Linneo). Cóndor, mainly of falcon family with talons. The largest of all winged creatures. Has darkish or black feather coat and the adults have a white ruff or collar. It feeds on carrion but when very hungry may attack live ones.
Lloq'e: Left, ritual complement in the distribution of space within the Andean Culture.
Paña: Right. Ritual complement in Andean construction designs.
Pachakamaq: Pacha: Earth, world, time; Kamaq: the maker, arranger, the one that controls. Supreme being, Lord of the Universe. The being that controls earthquakes and governs all things.
Pachamama: Mother Earth.
Qarpay: Spray, sprinkle.
Q'ente: Zoo. (Kolibrí coruscans gould and other species). Hummingbird, Kolibrí Bird of the family troquilidae, has a small body, bright green plumage, thin elongated beak; is also variable in form, length and colour. Capable of hovering flight.
Kuka: Qoqa: Bot. (Erythroxylon coca Lam.). (Etymol. Qoqa, Aymara term: divine plant or excellent plant). Has been utilized from the beginning of time for its nutritional virtues as well as its magical effects; was known as the spiritual food on people.
Qosqo: The navel, the center.
Salqantay: (Etim. Salqa-antay, surly, short-tempered, copper producing). Salqantay, important snow capped mountain of Willkamayu (Vilkanota) watershed; situated in the district of Mollepata, Province of Anta, department of Qosqo. Elevation: 6,263 meters above sea level. Regarded by the Inkas and present day natives as an Auki or Apu.
Sami: Luck, fortune, satisfaction, happiness.
Tupay: Encounter, interview, convergence and also competition.
Ukhu Pacha: The inner or lower world of caves, of openings into the earth and the underground; by extension: "the realm of the dead;" or the kingdom into which Pachamama receives to cleanse all our wrong doings or crimes.

Waka: Huaca, worship site, temple, sacred ground, place of devotion or ritual. By extension: place of the Sacred Spirit.
Willka: Sacred, Divine.
Willkamayu: Willkanota. Sacred River of the Inkas. Its source lies in the mountain area known as the Nudo de Vilkanota (Willkanota) in the zone of La Raya; the boundary between the Regions of Cusco and Puno, at an elevation of 5,846 meters above sea level. It flows through the Sacred Valley of the Inkas until entering the jungle region where its name changes at first to Alto Urubamba and later to Bajo Urubamba.
Wiraqocha: Supreme God and Deity of the Andes. Creator of all human reality. Lord of Heaven and Earth. Is represented by a Solar Being holding two staffs.
Yanantin: A Pair, or couple of lovers. By extension the name is also given to certain mountains that have twin peaks or appear to be two mountains united into one, as if they represented a couple.

NOTE: The basis for definitions of some Quechua words were taken from the Quechua-Spanish-Quechua Dictionary of the Main Academy of Quechua Language (Qosqo, Peru, 1995). Other definitions have been, with great respect, literally adapted for this work with the goal of amplifying their symbolic meaning.

BIBLIOGRAPHY

ACADEMIA MAYOR DE LA LENGUA QUECHUA
1995 Diccionario Quechua-Español-Quechua. Municipalidad - Qosqo.
AGURTO CALVO, Santiago
1987 Construcción, Arquitectura y Planeamiento Incas. Lima - Perú. Cámara Peruana de la Construcción.
ANGLES VARGAS, Víctor
1998 Machu Picchu y el Camino Inca. Lima - Perú. INDUST. gráfica.
BENTLY, Peter
1993 World Mythology. Duncan Baird Publishers.
BINGHAM, Hiram
1949 Machu Picchu, la ciudad perdida de los incas. Santiago de Chile. Empresa editora Zig-Zag, S.A.
BOLLINGER, Armin
1996 Así construían los inkas. La Paz. Editorial "Los Amigos del Libro".
CLEAVE, Andrew
1995 Big Cats, a portrait of the animal world. New York. Todtri Productions Limited.
GASPARINI, Graciano/MARGOLIES, Luise
1977 Arquitectura Inka. Caracas. Universidad Central de Venezuela.
GOUGAUD, Henri
1995 Les sept plumes de l´aigle. Francia. Editions du Seuil.
GRANDES NARRACIONES
Historiadores de Indias. España. TXERTOA EDICIONES.
GRIFFITHS, Nicholas
1997 La Cruz y la Serpiente. Pontificia Universidad Católica del Perú.
KAUFFMANN, Federico
2001 SEX IN ANCIENT PERÚ. Quebecor World Perú S.A. Lima.
JULIEN, Nadia
1992 Enciclopedia de los Mitos. Marabout, Belgique.
MASON, J. Alden
1961 Las Antiguas Culturas del Perú. México. Fondo de cultura económ.
MCINTYRE, Loren
1975 The incredible Incas and their timeless land. The National Geographic Society. Washington, D.C.
MIDDENDORF, Ernst
1974 PERÚ. Dirección Universitaria, Biblioteca y Publicaciones de la Universidad Nacional Mayor de San Marcos. Lima.
MITCHELL, Ruby
1991 A virgem do Sol, os últimos incas. Sao Paulo, Br. Círculo do livro.
MUJICA B., Elías
1992 Seminario-Taller Internacional Arqueología del Santuario Histórico Nacional y Sitio Patrimonio Mundial de Machu Picchu. Cusco.

PROTZEN, Jean-Pierre
1993 Inca Architecture. New York. OXFORD UNIVERSITY PRESS.
RUIZ DURAND, Jesús
2004 Introducción a la iconografía andina 1. IDESI/BID. Lima.
SZEMINSKI, Jan
1996 Wira Quchan y sus obras. Lima - Perú. IEP ediciones.
UHLE, Max
1997 Max Uhle y el Perú Antiguo. Pontificia Universidad Católica del Perú.
VALENCIA ZEGARRA, Alfredo/GIBAJA OVIEDO, Arminda
1992 Machu Picchu la Investigación y Conservación del Monumento Arqueológico después de Hiram Bingham. Qosqo - Perú. Municipalidad del Qosqo.
VIDAL, Humberto
1958 Visión del Cusco. Cusco - Perú. Editorial Garcilazo.
VON HAGEN, Adriana/MORRIS, Craig
1998 The Cities of the Ancient Andes. Hong Kong. Thames and Hudson Ltd, London.
WIENER, Charles
1993 PERÚ Y BOLIVIA. Instituto Francés de Estudios Andinos. Universidad Nacional Mayor de San Marcos.

GENERAL INDEX